NATO ASI Series

Advanced Science Institutes Series

A series presenting the results of activities sponsored by the NATO Science Committee, which aims at the dissemination of advanced scientific and technological knowledge, with a view to strengthening links between scientific communities.

The Series is published by an international board of publishers in conjunction with the NATO Scientific Affairs Division

A Life Sciences **B Physics**	Plenum Publishing Corporation London and New York
C Mathematical and Physical Sciences **D Behavioural and Social Sciences** **E Applied Sciences**	Kluwer Academic Publishers Dordrecht, Boston and London
F Computer and Systems Sciences **G Ecological Sciences** **H Cell Biology** **I Global Environmental Change**	Springer-Verlag Berlin Heidelberg New York London Paris Tokyo Hong Kong Barcelona Budapest

PARTNERSHIP SUB-SERIES

1. Disarmament Technologies	Kluwer Academic Publishers
2. Environment	Springer-Verlag/Kluwer Academic Publishers
3. High Technology	Kluwer Academic Publishers
4. Science and Technology Policy	Kluwer Academic Publishers
5. Computer Networking	Kluwer Academic Publishers

The Partnership Sub-Series incorporates activities undertaken in collaboration with NATO's Cooperation Partners, the countries of the CIS and Central and Eastern Europe, in Priority Areas of concern to those countries.

NATO-PCO DATABASE

The electronic index to the NATO ASI Series provides full bibliographical references (with keywords and/or abstracts) to about 50000 contributions from international scientists published in all sections of the NATO ASI Series. Access to the NATO-PCO DATABASE compiled by the NATO Publication Coordination Office is possible in two ways:

- via online FILE 128 (NATO-PCO DATABASE) hosted by ESRIN, Via Galileo Galilei, I-00044 Frascati, Italy.

- via CD-ROM "NATO Science & Technology Disk" with user-friendly retrieval software in English, French and German (© WTV GmbH and DATAWARE Technologies Inc. 1992).

The CD-ROM can be ordered through any member of the Board of Publishers or through NATO-PCO, Overijse, Belgium.

2. Environment – Vol. 14

Springer
Berlin
Heidelberg
New York
Barcelona
Budapest
Hong Kong
London
Milan
Paris
Santa Clara
Singapore
Tokyo

Economics of Atmospheric Pollution

Edited by

Ekko C. van Ierland

Department of General Economics
Wageningen Agricultural University, P.O. Box 8130
6700 EW Wageningen, The Netherlands

Kazimierz Górka

Department of Industrial and Environmental Policy
The Cracow Academy of Economics
ul. Rackowicka 27, room 155
31-150 Kraków, Poland

Springer

Published in cooperation with NATO Scientific Affairs Division

Selected papers presented at the NATO Advanced Research Workshop "The Economics of Atmospheric Pollution, Theories, Applied Models and Implications for International Policy Making", held in Wageningen, The Netherlands, November 16–18, 1994

Library of Congress Cataloging-in-Publication Data

Economics of atmospheric pollution / edited by Ekko C. van Ierland.
p. cm. -- (NATO ASI series. Partnership sub-series 2, Environment ; vol. 14)
"Proceedings of the NATO Advanced Research Workshop "The Economics of Atmospheric Pollution, Theories, Applied Models and Implications for International Policy Making", held in Wageningen, The Netherlands, November 16-18, 1994"--T.p. verso.
"Published in cooperation with NATO Scientific Affairs Division."
Includes bibliographical references and index.

1. Climatic changes--Economic aspects--Congresses. 2. Greenhouse gases--Economic aspects--Congresses. 3. Air--Pollution--Economic aspects--Congresses. 4. Global warming--Economic aspects--Congresses. I. Ierland, E. van (Ekko van) II. Górka, Kazimierz. III. North Atlantic Treaty Organization. Scientific Affairs Division. IV. NATO Advanced Research Workshop "The Economics of Atmospheric Pollution, Theories, Applied Models and Implications for International Policy Making" (1996 : Wageningen, Netherlands) V. Series.
QC981.8.C5E247 1996
363.73'922--dc20 96-30955
CIP

ISBN-13: 978-3-642-64744-4 e-ISBN-13: 978-3-642-61198-8
DOI: 10.1007/978-3-642-61198-8

Softcover reprint of the hardcover 1st edition 1996

Typesetting: Camera-ready by authors/editors
SPIN: 10492445 31/3137 – 5 4 3 2 1 0 – Printed on acid-free paper

Preface

This book contains a selection of papers that have been prepared for the NATO Advanced Research Workshop on the Economics of Atmospheric Pollution, that took place in Wageningen, The Netherlands, November 1994, hosted by Wageningen Agricultural University and sponsored by NATO Scientific and Environmental Affairs Division. Participants from the USA and a large number of countries in Western, Central and Eastern Europe have participated to discuss the economic aspects of transboundary air pollution and climate change.

A number of selected papers have been reviewed and revised on the basis of the comments provided. The editors kindly acknowledge the support of Prof. Charles Kolstad, University of California, Santa Barbara, and Prof. Stef Proost, Center for Economic Studies, Catholic University Leuven, for reviewing several chapters of the book. Also the assistance of several anonymous reviewers is kindly acknowledged.

We hope that the book will contribute to a better understanding of the most relevant issues in the area of international policymaking on transboundary pollution and climate change, and that it contributes to further economic analysis in this interesting research area. The topic of transboundary pollution related to climate change, acidification and tropospheric ozone will in the coming decades continue to be relevant for all countries in the world.

Ekko van Ierland

Kazimierz Górka

Wageningen/Cracow, June 1996

CONTENTS

1 On the Economics of Atmospheric Pollution

Ekko van Ierland
Department of General Economics
Wageningen Agricultural University
P.O. Box 8130
6700 EW Wageningen
The Netherlands

Abstract

This chapter provides an introduction on the economics of atmospheric by discussing the various aspects that are related to the origins of atmospheric pollution, the analysis and the possible policy instruments for solving the problems. The chapter describes how atmospheric pollution has developed from the local level to the global level and how uniformly mixing pollutants and non-uniformly mixing pollutants are involved in it, each requiring its specific policy approaches. The chapter concludes with a brief overview of the various chapters of the book.

1.1 Introduction

The atmosphere and its gases are essential for all aspects of life on earth. Clean air, clear visibility, proper protection by the ozone layer against harmful radiation and an acceptable level of radiative forcing by the natural greenhouse effect are essential ingredients for sustainable development of societies in North, South, East and West. The physical and chemical characteristics of the atmosphere are complex and to some extent only partially understood by natural scientist and climatologists. By its very nature, climate is difficult to analyze and predict which makes the economic analysis of atmospheric pollution not only a complicated, but also a challenging task.

NATO ASI Series, Partnership Sub-Series, 2. Environment – Vol. 14
Economics of Atmospheric Pollution
Edited by Ekko C. van Ierland and Kazimierz Górka

Of course, economists would like to present well-designed cost benefit studies for the various aspects of atmospheric pollution, in order to provide guidance and advise to national and international policy makers. In practice a full cost benefit analysis is not easy - if at all - to establish. For the costs of emission reduction fairly well documented and reliable technical costs estimates are available, for example documented in the RAINS model for acidification in Europe (Alcamo et al, 1990) or in the IPCC second assessment report (IPCC, 1995) as far as climate change is concerned. However the damage estimates for atmospheric pollution are extremely difficult to make, since a large variety of non-market values are included, such as existence value of ecosystems and species, bequest value and option value. Although a broad range of valuation methods is advocated in the literature, including travel cost method, averting behaviour method, contingent valuation method and various hedonic pricing methods, it is generally accepted that no profound cost benefit analysis of atmospheric pollution is possible, that includes all the damage categories in monetary units.

For these reasons a large number of integrated assessment models have been developed for identifying efficient and effective policy options in all fields of atmospheric pollution. Some of the models include damage assessments, others are restricted to effects in physical and ecological terms, or they restrict them selves to prespecified constraints in terms of concentrations or critical loads. At present most of these models are problem oriented and focusing on a single issue in atmospheric pollution, for example acidification, global warming or tropospheric ozone. Most models provide a partial analysis and do not yet focus on joint emission reduction for various pollutants in various areas of atmospheric pollution. Since synergistic between pollutants exist and since most pollutants are jointly related to the combustion of fossil fuels (or can jointly be abated by the introduction of renewables) a need exists for combined integrated analysis of various aspects of atmospheric pollution. Until now only a few studies (for example Zylicz, 1994; Heyes et al., 1995;) are focusing on these interactions and this area is certainly very stimulating for further research.

1.2 Background of atmospheric pollution

Environmental problems related to atmospheric pollution are numerous and can be categorized as follows according to the regional scale:

(i) local air pollution, like smog in cities and pollution near factories;

(ii) pollution at the regional level, like acidification and formation of tropospheric ozone;

(iii) pollution at the global level, like the enhanced greenhouse effect, as a result of increasing concentrations of greenhouse gases (e.g. CO_2, CH_4, CFC's and N_2O) or ozone layer depletion by CFC's.

Another classification concerns the sources of pollution:

(i) point sources of pollution (mainly from stacks);

(ii) non-point sources of pollution, like methane emissions from rice field or emissions from non-stationary sources in traffic;

Table 1.1. Contribution of compounds to categories of atmospheric pollution.

compound	acidification	photochemical smog	global warming	ozone depletion
CO_2			direct	
CO		indirect		
CH_4		indirect	direct	
C_xH_y		direct		
NO_x	direct	direct		
N_2O			direct	
NH_3/NH_4^+	direct			
SO_2	direct		negative	
CFC			direct	direct
O_3		direct	direct	

Source: Graedel and Crutzen (1993)

Some of the emissions are directly related to the combustion of fossil fuels, others are typical process emissions that are related to industry or agriculture. Apart from anthropogenic sources, the natural sources and sinks play a role.

Several compounds have a key influence on more than one aspect of atmospheric pollution. For example NO_x plays a role in acidification, together with SO_2 and NH_3. At the same time NO_x is a precursor of tropospheric ozone, together with VOC's. By itself NO_x contributes to eutrophication, which requires reduction of NO_x, regardless of its acidifying characteristics. To complicate matters even more, it is also observed that SO_2 not only contributes to acidification, but may have a cooling effect as far as global warming is concerned. Finally, now soil and water pollution from point sources are more and more restricted, the relative contribution of atmospheric pollution as a source of soil and water pollution is increasing in many areas of the world, particularly for persistent pollutants and heavy metals. The main effects of the various compounds on problems of atmospheric pollution are summarized in table 1.1.

1.3 Economic problems related to atmospheric pollution

The economic analysis of atmospheric pollution has a rather long history, that goes back to the early examples of air pollution in the industrial areas and cities in the beginning of the industrial revolution. Examples of external effects at the local level have led to the development of the theory of externalities in welfare theory and environmental economics, associated with the names of Marshall and Pigou. Nobel Prize laureate Coase (1960) presented an elegant analysis of the impact of property rights on the optimal emission level, and noted that under specific circumstances, such as absence of transactions costs, the optimal level could be reached (where marginal costs equate marginal benefits) by negotiations between polluter and pollutee, regardless of the structure of the property rights. Early work by Tietenberg (1985) and Baumol and Oates (1988) has asked attention for the various policy instruments that can be applied to solve local and transboundary pollution, both for uniformly mixing pollutants and non-uniformly mixing pollutants. The central question in the analysis is how to reach efficient and effective solutions that are acceptable for national and international policy makers, and that can be applied in practice. The systems should avoid excessive transactions costs, should be easy to monitor and compliance should be guaranteed. Where these requirements are difficult to meet for national environmental problems, they are even more complicated in an international context, where losers and winners in international agreements have different interest and where the incentive to cheat and free ride are numerous.

The topic is extended from local air pollution to the analysis of acidification (a typical example of non-uniformly mixing pollutants), and to the analysis of global warming due to greenhouse gases (typical examples of uniformly mixing pollutants), where the location of emissions is irrelevant for the environmental impacts. Some studies pay attention to the synergistic effects of various pollutant, where the environmental impacts are non-linearly related to the various levels of emissions, because of chemical interactions between several pollutants. This is typically the case with tropospheric ozone where ozone is formed from its precursors NO_x and VOC in a typically non-linear way (cf. Kelly and Gunst, 1990).

1.4 Policy instruments

Nowadays, both in the literature and in practice, a wide range of policy instruments can be found aiming at the reduction of air pollution policy instruments. These range from (i) command and control policies to (ii) emission charges and (iii) tradable discharge permits, where categories ii and iii are indicated as economic instruments.

(i) Command and control policies are widely used and include the setting of ambient standards, product specifications (catalysts on cars) or process specification, for example the European guidelines for large combustion installations in Europe.

(ii) Emission charges are gradually introduced in various countries in Europe and are sometimes implicit in existing excise taxes on various types of fuels. The carbon tax proposals are widely discussed, but not yet implemented, with a few exceptions and to a limited extent in the smaller countries in Europe (Scandinavian countries and the Netherlands). In the literature not only emission taxes are discussed, but also charges on the export of pollutants to other countries, as calculated by atmospheric dispersion models (Mäler, 1989).

(iii) Tradable discharge permits are widely discussed in the literature (cf. Tietenberg, 1994). In practice they are used in the USA for sulphur emissions and to a limited extent for the prevention of smog in city centres.

Tradable discharge permits can be 'grandfathered', i.e. distributed by the government to enterprises on the basis of historical emissions (or an other relevant distribution mechanism), or they can be auctioned according to various bidding schemes. For uniformly mixing pollutants tradable permits can be an efficient and effective policy instruments, explaining why it is often advocated for application in global warming policies. For non-uniformly mixing pollutants the system is problematic because the location of the polluting sources are not considered. This may lead to a concentration of sources of emissions in a specific area, such that local ambient air quality standards are exceeded or that excessive damage to vulnerable ecosystems occurs. To solve for these problems several methods are advocated in the literature for example a system in which permits are traded on the basis of an exchange rate unequal to one (depending on the locations of the sellers and buyers), implying that the regional distribution is considered (Klaassen, 1995).

Another approach is to set additional restrictions to the trading of permits, for example set upper limits to the quantities that may be traded by each buyer and seller, such that a cost effective solution is reached for achieving prespecified deposition targets at the various receptor points (Kruitwagen, Hendrix and Van Ierland, 1994). In addition to this type of analysis attention is paid to banking of emission permits, offset systems and the future market for tradable discharge permits.

Air pollution policies are closely related to energy demand and supply policies that by affecting the level of energy consumption and changing the fuel mix have also an important impact on the emissions of energy related air pollutants. For this reason the various aspects of air pollution are often included in energy supply models, like the MARKAL model. In these energy optimization models the choice of energy technologies is not only dependent on the cost of the various technologies, but also on their contribution to emissions of pollutants. By setting constraints on the level of emissions for the various pollutants, scenarios can be provided for energy supply technologies that may cost effectively contribute to reaching the prespecified emission targets (Kram, 1994).

Policy measures for atmospheric pollution will be different in various areas of the world, depending on the local circumstances, the resources available and the level of development. In the third world air pollution is rapidly increasing because of fast economic development, particularly in the far East and China, where coal is the dominant source of primary energy. Energy efficiency in these countries and in Central and Eastern Europe is still rather low as compared to industrialised countries, due to the existing capital stock and the level of technology embodied in it. This implies that technological progress in the field of energy efficiency improvement, dispersion of technological know how and the introduction of cleaner and sustainable technologies are essential ingredients for successful policies in the area of atmospheric pollution, apart from the application of end of pipe technologies such as scrubbers and catalysts.

A special topic in atmospheric pollution is the danger of industrial accidents and explosions, like the disaster in Bhopal or the nuclear accident in Chernobyl. These require strict safety standards and the application of modern and up to date technologies, in combination with good management. These aspects that have many economic aspects, however, are explicitly not discussed in this book.

1.5 Modeling and methodologies

Environmental economic analysis of atmospheric pollution is largely based on modeling the relations between economic activity, energy consumption, emissions of compounds, the dispersion in the atmosphere, the resulting concentrations and/or depositions at various receptor points and their effects. Together with the costs associated with the reduction of emissions (either by energy conservation, end of pipe reduction, or the creation of additional sinks in the case of greenhouse gases) this is the set up for the cost benefit analysis, or the cost efficiency analysis on which policy recommendations are based. The analysis in the field of non-uniformly mixing pollutants (like acidification and tropospheric ozone) traditionally started with the straightforward cost benefit and cost efficiency studies as specified in environmental economic textbooks like Baumol and Oates (1988) and Tietenberg (1992). The general framework is to minimize the sum of abatement cost and damage costs, given the abatement cost functions, the damage cost function and the source-receptor matrix that describes how the pollutants are dispersed in the atmosphere. This type of analysis provides a cost efficient solution, provided that all countries or actors are willing to cooperate and that they are willing to contribute their share in the abatement costs. Eventually a system of side payments can be used that makes sure that all countries are better off by cooperating than by not cooperating (cf. Mäler, 1989).

A different strand of analysis focuses on the strategic position of the various actors and there willingness or unwillingness to participate in the various coalitions that are possible. This analysis is based on game theory and provides insight in the strategic behaviour of countries and the likelihood of reaching international cooperation. Given the complexity of the analysis these models mainly focus on the theoretical aspects and pay less attention to the detailed empirical elaboration.

For the analysis of global warming a large number of integrated asessment models have been developed, following the early work of Nordhaus on the DICE model (Nordhaus, 1992) and the regionalized version of it (Nordhaus, 1995). This model includes a highly aggregated description of the development of the world economy based on a Keynes Ramsey growth model. This model is hardlinked to an highly simplified climate model that translates the emissions of greenhouse gases into greenhouse gas concentrations. From this the radiative forcing that is relvant for the change in temperature is derived. Next the change in average atmospheric temperature as compared to pre-industrial level is calculated. Finally the model contains a damage cost function which expresses the damage of global warming as a percentage of Gross world product. Estimates of monetary damages are shown in Table 1.2. In this way the model that maximizes discounted utitlity over a time period of about 100 years is able to provide scenarios in which the optimal level of emission reduction of greenhouse gases is calculated. A similar but more elaborated approach is followed by Manne and Richels in the MERGE model (Manne and Richels, 1995). These authors use a more detailed analysis of the world energy demand and supply system in order to calculate the emissions of carbon dioxide for five regions in the world. In addition the model includes other greenhouse gases and a rather detailed climate to calculate the change in average atmospheric temperature. The model can be used to analyse the efficiency of various policy options for reducing greenhouse gas emissions and to assess the optimal time path of emission reduction. The calculations illustrate that considerable cost savings can be ob-

Table 1.2. Climate change damage ($2\times CO_2$ $-10^9\$^{(1988)}$)

regions[a]	1	2	3	4	5	6	7	8	9	total
coastal defence	1.5	1.7	1.8	0.5	0.0	1.0	2.0	0.5	0.5	9.5
dryland loss	2.0	0.5	4.0	1.3	0.0	0.5	1.0	0.0	0.5	9.8
wetland loss	5.0	4.0	4.5	1.3	0.0	1.5	1.5	0.5	0.5	18.8
species loss	5.0	5.0	5.0	2.5	0.0	2.0	1.0	1.0	0.5	22.0
agriculture	10.0	-4.2	-6.3	-32.5	0.7	11.8	21.0	21.0	17.5	14.5
amenity	12.0	12.0	12.0	-1.0	0.1	0.4	1.2	1.2	0.5	38.0
life/morbidity[b]	37.7	36.6	36.9	18.8	0.6	11.4	21.6	15.4	0.0	188.0
migration	1.0	1.1	0.6	0.5	0.0	2.4	4.4	2.6	1.3	13.8
natural hazards	0.3	0.0	0.8	0.0	0.0	0.1	0.1	0.1	0.3	1.4
total	74.0	56.5	59.0	-7.9	1.3	31.0	53.6	18.0	30.3	315.7
(% GDP)	(1.5)	(1.3)	(2.8)	(-0.3)	(4.1)	(4.3)	(8.6)	(5.2)	(8.7)	(1.9)

a 1 = USA and Canada; 2 = OECD-Europe; 3 = Japan, Australia and New Zealand; 4 = Central and Eastern Europe and the former Soviet Union; 5 = the Middle East; 6 = Latin America; 7 = South and South East Asia; 8 = Centrally Planned Asia; 9 = Africa.

b The value of a human life is assumed to equal $250,000 plus 175 times the per capita income in the region.

Source: Tol *et al.* (1995)

tained if both regional and temporal optimization of emission reduction takes place. Joint implementation allows for regional optimization of emission abatement strategies. Careful analysis of the dynamic pattern of emmision reduction allows for optimization in time, taking into account the relevant discount factor (Manne and Richels, 1995). Similar models are developed to deal in an empirical setting with the topic of uncertainty and learning about climate change. Following the analysis of Kolstad (1994), several authors have elaborated on this topic, for example Peck and Teisberg (1993) or Ulph and Ulph, 1994. The analysis is further elaborated in chapter 2 and 3 of this volume.

1.6 Cooperation or free riding?

The main challenge for future policies with regard to atmospheric pollution is the question whether nations will be willing to cooperate to reduce the emissions of polluting and harmfull compounds including the greenhouse gases and if so under what conditions. The 1992 conference in Rio de Janeiro has shown how complicated it is to develop a common way of thinking in this area and how strong national interests are. The follow up conference in Berlin 1995 provided for further research, but did not yet result in clear policy recommendations and commitments to reduce emissions.

Theoretical and empirical analysis clearly shows that full cooperation offers the best opportunities for the set of nations as a whole. On the other hand each nation is facing strong incentives to free ride, to wait and see till other countries take actions. It is evident that strict greenhouse policies may be very expensive for individual countries to pursue, because the competitive position and the danger of carbon leakage when firms locate in other countries or when production in these countries is expanded. The possibility of countereffective policies has been mentioned in the literature where the imposition of a carbon tax in a single country may actually result in a net increase of carbon emissions at the world level, because of low energy efficiency in competing countries.

The main question will be who is willing to bare the costs of emission reduction of greenhouse gases and which intstruments the international community has available to convince other countries to participate in a coalition for the reduction of the emissions of greenhouse gases. The social pressure factor or trade barriers have been mentioned as options to stimulate participation and to punish countries that act as free riders (Hoel and Schneider, 1995). Clearly the question concerns not only the distribution of the costs of emission reduction between countries, but also the distribution of the costs and benefits between the present and future generations. If the greenhouse problem is a serious problem the danger exists that the costs will be transferred to next generations. If they are not sufficiently compensated by other means (capital accumulation, investment in education and technology), this clearly is not an example of what is considered 'sustainable development'. Clearly, the economics of atmospheric pollution will play an important role in the political relations between North and South and between East and West in the coming decades. This calls for further research and analysis on international cooperation to protect the earth's atmosphere.

1.7 The structure of the book

The book starts with two chapters on uncertainty and learning in the case of global warming. Chapter 2, by Charles Kolstad, deals with uncertainty and learning in the case of global warming, taking into account stock externalities and capital irreversibilities. The chapter contains a challenging analysis on the question whether the threat of increasing atmospheric temperature requires immediate action or whether future learning and an attempt to avoid irreversible investment in greenhouse gas emission reduction allows a postponement of reduction of emissions of greenhouse gases till later. On the other hand policies should aim at avoiding irreversible climate change as a result of the accumulation of a stock pollutant. The chapter deals with the key questions of economics of how to make decisions under uncertainty, and how to reduce uncertainty by investing in more information and learning.

Kolstad concludes that unless we learn about climate change such that we might wish to negatively emit in the future (to reduce the stock of greenhouse gases), there should be no tendency to under emit today to forestall irreversible environmental effects. If we wish to reduce risk by restricting emissions today to avoid low-probability catastrophic effects in the future, this bias in emission control is due to risk aversion, not the irreversibility effect.

Chapter 3, by Alistair and David Ulph, continues on the same topic, but integrates the analysis of strategic interactions between independent national governments that should take decision on their policies with regard to global warming. In a bright and clear analysis Ulph and Ulph show that careful microeconomic analysis is capable of providing in-depth insight in the relevant factors that are determining the optimal solution for policies on greenhouse gases emission reduction.

The results are very interesting and challenging, since Ulph and Ulph conclude that in the new setting learning in the future is likely to cause current emissions to be lower than they would be without learning and this can be true of individual countries or of all countries. It is also illustrated that there are greater gains from cooperation when there is the possibility of learning.

Chapter 4, by Johan Eyckmans and Stef Proost, focuses on voluntary agreements for greenhouse gas emission reduction policies, starting from the Nash Cournot equilibrium. Countries are then provided an opportunity to buy or sell additional emission reduction at a fixed price in an international abatement market. It is shown that the country with the highest willingness to pay for environmental quality determines the total level of emission reduction. It is illustrated that in some cases, in a setting of monopsonistic trade, it is possible to achieve a Pareto improvement compared to the Nash Cournot equilibrium without trade. A necessary, but not a sufficient condition would be that the monopsonist applies some form of price discrimination in the international abatement market.

Chapter 5, by Ronnie Schöb, provides some surprising results of policy schemes of the green tax. In stead of providing a double dividend, as suggested by some authors, the paper show that a green tax may even lead to an increase in carbon emissions in an empirically relevant range of parameters. In the case of revenue neutral green tax reform, tax rate cuts for at least one non-polluting good can occur, potentially resulting in an increase of carbon emissions, because of the complementarity/substitutability relationships which might exist between clean goods and polluting goods. The model includes two generations of households and analyses the first best and second best solutions. To illustrate the analysis an empirical application is provided based on the German context with regard to car traffic, in which it is illustrated that a price elasticity of -0.3 for the clean good is sufficient to obtain the paradoxical case. The paper clearly illustrates that in making decision on the green tax we should be aware of the effects of the tax and of the effects of the accompanying policy measures.

Chapter 6, by Stephen Peck and Steve Wan, analyzes for the case of global warming, the policy impacts of various climate damage functions, utilizing simple greenhouse gas emission reduction cost models. The same model is used to evaluate the value of information for key parameters. The analysis is based on a simple optimal control model, assuming a linear damage function and a stepped damage function. In the latter case the control strategy is to keep the incremental greenhouse gas mass at or below the threshold value of the step function. In addition it is analysed by way of examples what the value is of information that reduces the uncertainty on specific parameter values in the model.

Chapter 7, by Yuri Ermoliev, Andries Nentjes and Ger Klaassen, provides a challenging design of a cost effective ambient charge under incomplete information and risk, in the context of a non-uniformly mixing pollutant, for example the case of acid rain. One of the challenging questions in the case of non-uniformly mixing pollutants is the question of how to design a first best policy regime for reaching deposition targets, in a situation where the policy maker has no or only limited information on the emission control costs of various actors. This question is in practice not yet solved, nor in the USA nor in Europe. The system of tradable discharge permits for sulphur emissions in the USA does not take into account the location of the emissions and may result in an undesired regional pattern of emissions with high levels of deposition at receptors with vulnerable and sensitive ecosystems, characterised by low critical loads. In Europe, the second sulphur protocol is based on calculations with the RAINS model, that optimizes the distribution of emission reduction over various countries, on the assumption that full information on the actual emission control costs is available to the policy maker or its advisors. What, however, if the used information on emission control costs is incorrect, maybe partially because of strategic behaviour of the reporting countries?

The chapter analyses how a system of emission charges can be applied that results by a process of trial and error in the cost efficient solution for reaching prespecified environmental targets in terms of critical loads at various receptor points. For this purpose an excess pollution function is designed that is defined as the gap between actual concentrations and the targets concentrations (ambient standards) at receptors. The system should be facilitated by a network of computers that provides immediate information from the actors to the agencies and vice versa. The authors argue that the system, which functions as a kind of market by trial and error, can result in convergence to the cost effective solution. In the initial deterministic analysis it is assumed that the dispersion matrix of pollutants is known (the source-receptor matrix is known to all actors and the policy makers). In the second part of the analysis the transfer coefficients are random.

The final chapter, Chapter 8, by Inge Mayeres and Stef Proost, provides an overview of various aspects of the economic analysis of tropospheric ozone abatement in Europe, on the basis of an abstract model with growing complexity. The paper shows how tropospheric ozone is formed from its precursors NO_x and VOC that are transported over long distances. NO_x also plays a role in acidification. Next the analysis is extended by introducing transfrontier pollution and several other characteristics like stochastic pollution, defensive expenditures and the properties of the abatement cost functions. The analysis focuses on the damages to health and agriculture and the emission reduction costs for the two precursors. The paper concludes that further environmental economic modelling is necessary for developing a cost effective and efficient ozone abatement strategy in Europe.

1.8 Conclusions

The economics of atmospheric pollution is a very interesting area of research in environmental economics. Unfortunately the problems of atmospheric pollution have by no means been solved until now, and continuing economic growth and related energy consumption, both in the industrialised and the developing world will continue to result in increasing emissions of various pollutants and greenhouse gases. Although in some areas of the world progress is being made in reducing some of the harmful emissions, in most other areas the annual levels of emissions are increasing rapidly. Unless appropriate policies are pursued and implemented it can be expected that almost all problems that have been mentioned at the beginning of the chapter will continue to deteriorate, may be with the exception of acidification in Europe and the USA, and the depletion of the ozone layer, which seems to recover from the stress that it has been exposed to now emissions of CFC are being reduced.

It is a challenging task for environmental economists to continue the analysis of atmospheric pollution, in cooperation with atmospheric chemists and meteorologists, in order to identify optimal policies, that are based on sound scientific analysis, both of the natural sciences aspects and the socioeconomic aspects of these problems. The analysis should not only focus on the fundamental research required to understand the basic fundamentals of the processes at hand, but also on the application in actual policy making in order to achieve the results that are necessary for sustainable development and the safeguarding of the earth's atmosphere. It definitely is a challenge to provide the analysis that is required to provide support for international policy making, not only at local and regional level but also on the global level. Acidification policies in Europe show that it is possible, at least to some extent, to reach international agreement on emission reduction, which goes beyond the Nash Cournot equilibrium as described by Mäler. Whether in the long run the same can be noticed on international policies for global warming cannot be observed by now. At least we can hope that economic analysis can contribute to the efficiency and effectiveness of proposed policy solutions and policy regimes.

References

Alcamo, J., R. Shaw and L. Hordijk (EDS.) (1990) *The RAINS Model of Acidification. Science and Strategies in Europe*. Kluwer Academic Publishers, Dordrecht.

Baumol, W.J. and W.E. Oates (1988) *The Theory of Environmental Policy*, Second Edition, Cambridge University Press, Cambridge.

Coase, R.H. (1960) The Problem of Social Cost, *Journal of Law and Economics*, October, pp.1-44.

Graedel, T.E. and P.J. Crutzen (1993) *Atmospheric Change - An Earth System Perspective*, W.H. Freeman and Company, New York.

Heyes, C., W. Schöpp and M. Amann (1995) *A Simplified Model to Predict Long Term Ozone Concentrations in Europe*, International Institute for Applied System Analysis, Laxenburg, Austria.

Hoel, M. and S.H. Schneider (1995) *Incentives to participate in an international environmental agreement*, Paper presented at the sixth annual conference of the EAERE, Umeo, Sweden.

IPCC (1995) *The Second Assessment Report of Working Group II*, Cambridge University Press, Cambridge.

Kelly, N.A. and R. Gunst (1990) Response of Ozone to Changes in Hydrocarbon and Nitrogen Oxide Concentrations in Outdoor Smog Chambers filled with Los Angeles Air, *Atmospheric Environment*, vol. 24A, no. 12 pp. 2991-3005.

Klaassen, G. (1995) *Trading Sulphur Emission Permits in Europe Using an Exchange Rate*, in E.C. van Ierland (ed.) International Environmental Economics - Theories, Models and Applications to Climate Change, International Trade and Acidification, Developments in International Economics-4, vol.4, Elsevier Science B.V., Amsterdam.

Kruitwagen, S., E. Hendrix and E. van Ierland (1994) *Tradeable SO_2 permits: Guided Bilateral Trade in Europe*, in E.C. van Ierland (ed.) International Environmental Economics - Theories, Models and Applications to Climate Change, International Trade and Acidification, Developments in International Economics-4, vol.4, Elsevier Science B.V., Amsterdam.

Mäler, K-G. (1989) *The Acid Rain Game*, in H. Folmer and E. van Ierland (ed.) Valuation methods and policy making in environmental economics, Studies in Environmental Science 36, Elsevier.

Nordhaus, W.D. (1992) An Optimal Transition Path for Slowing Climate Change, *Science*, November.

Nordhaus, W.D. and Z. Yang (1995) *RICE: A Regional Dynamic General Equilibrium Model of Optimal Climate-Change Policy*, Yale University and MIT.

Peck, S.C. and T.J. Teisberg (1993) Global Warming Uncertainties and the Value of Information: An Analysis Using CETA, *Resource and Energy Economics* 15, 71-97.

Tietenberg, T.H. (1985) *Emission Trading - An Exercise in Reforming Pollution Policy*, Resources for the Future, Inc. Washington D.C., pp 222.

Tietenberg, T.H. (1992) *Environmental and natural resource economics,* Harper Collins Publishers, New York, pp 678.

Tietenberg, (1994) *Environmental Economics and Policy*, Harper Collins Publishers, New York, pp. 432.

Tol, R.S.J., T. van der Burg, H.M.A. Jansen and H. Verbruggen (1995) *The Climate-Fund - Some notions on the socio-economic impacts of greenhouse gas emissions and emission reduction in an international context*, Environmental Institute of the Free University Amsterdam.

Ulph, A. and D. Ulph (1994) *Global Warming: Why Irreversibilities May Not Require Lower Current Emissions of Greenhouse Gases,* Discussion Papers in Economics and Econometrics, No. 9402, University of Southampton.

Zylicz, T. (1994) *Improving Environment through Permit Trading: The Limits to a Market Approach*, in E.C. van Ierland (ed.) International Environmental Economics - Theories, Models and Applications to Climate Change, International Trade and Acidification, Developments in International Economics-4, vol.4, Elsevier Science B.V., Amsterdam.

2 Uncertainty, Learning, Stock Externalities and Capital Irreversibilities*

Charles D. Kolstad
Department of Economics
University of California at Santa Barbara
Santa Barbara
CA 93106, USA

Abstract

This paper concerns the irreversibilitity effect in stock externalities. In an environment of uncertainty with learning taking place, one may wish to under-emit today to avoid potential environmental irreversibilities. Alternatively, one may wish to under-invest in pollution control capital, avoiding investments in sunk capital that turn out to be wasted. The paper reviews theoretical results on the tension between these two effects and separates risk aversion from the irreversibility effect. The paper also presents a simple example of climate change policy.

2.1 Introduction

Uncertainty is a fundamental property of environmental externalities. Usually we understand poorly both the effects of these externalities and the costs of controlling them. This is one reason considerable sums are expended in trying to better understand environmental problems. Examples abound: hazardous wastes and ground water, global warming, acid rain, species extinction, pesticide accumulation, and the list could go on. An additional factor that

* Research supported by NSF Grant SBR-94-96303 and DOE Grant DE-FG03-94ER61944. Work conducted in part while the author was visiting the Catholic University of Leuven in Belgium and in part while on the faculty of the University of Illinois at Urbana-Champaign. Parts of this paper appeared in Kolstad (1996).

NATO ASI Series, Partnership Sub-Series, 2. Environment – Vol. 14
Economics of Atmospheric Pollution
Edited by Ekko C. van Ierland and Kazimierz Górka

frequently comes into play has to do with the cumulative or stock effects of the externality. For example, it is not the emissions of greenhouse gases that directly cause adverse effects; rather it is the stock of these gases that may lead to climate change. These two aspects of the problem -stock effects and uncertainty- lead to a tension between instituting control and delaying control. Some in society desire control of greenhouse gases before climate change is well understood, avoiding a difficult-to-reverse buildup of greenhouse gases. Others in society urge delaying control until the problem is clearly delineated, thus avoiding wasting control capital. If, *ex post*, the problem turns out to be less severe than expected, then those urging delay will have been proved correct (*ex post*). If on the other hand, the problem turns out to be more severe than expected, then delay can be very costly indeed.

The problem is that the future is uncertain and the potential risks are great. This problem is usually decomposed into two effects. One is risk aversion. Suppose we are contemplating an action today that will have an uncertain effect in the future. If one is risk averse, then the prospect of large effects in the future makes it worthwhile to pay a premium today (by taking otherwise suboptimal actions) to reduce the potential risks in the future. In the environmental economics literature this is usually called the option value of deferring a risky action, or how much one would be willing to pay to eliminate uncertainty regarding outcomes (Graham, 1981). This stands in contrast to a closely related effect pertaining to learning and irreversibilities, termed the quasi-option value (see Hanemann, 1989). When today's actions restrict tomorrow's opportunities, then we may want to moderate our choices today to keep options open tomorrow. The value of keeping options open in this context is the quasi-option value. Implicit in this is that tomorrow we will know more than today; otherwise we don't need options tomorrow. Consequently, the quasi-option value is the value of taking steps that maintain or increase tomorrow's choices. Learning must occur for there to be any value. Further, risk aversion is not necessary for there to be quasi-option value.

In the context of a stock externality, there are three types of irreversibilities that potentially may be involved, two on the damage side, and one on the control side. One irreversibility is climate catastrophes. Some suggest that an ice age or disappearance of the Gulf Stream could be triggered with no warning and little possbility of reversal of the effect. Another irreversibility on the damage side is that if one over-emits pollution and then finds the damage from the pollution stock is too high, one cannot immediately reduce that stock. Analagously, on the control side, if one invests in pollution control capital and then learns that damage is low, one cannot instantly reduce the abatement capital stock. To simplify, we do not consider climate catastrophes here -marginal damage from increased emissions is always finite. Thus the question is, do "irreversibilities" influence today's choice of emissions and if they do matter, what governs which is the dominant irreversibility -damage and control costs? Furthermore, it is important to distinguish between the influence on today's decision on how much to control originating in risk aversion vs. the influence due to the irreversibility effect (i.e., changing today's actions because they restrict tomorrow's options).

The existing literature is not totally definitive on the question of how learning and irreversibilities influence today's actions. The standard result in the literature is that when an environmental irreversibility is involved, one should bias action in favour of the environment (Henry, 1974; Arrow and Fisher, 1974; Freixas and Laffont, 1984). Epstein (1980) and later Ulph and Ulph (1994) have shown that this is not always the case. But this literature does not take into account the countervailing effect of the capital investment irreversibility (see Arrow, 1968). Furthermore, there is the literature on certainty equivalence that says, subject to some qualifications, that when today's actions only change tomorrow's costs, rather than restricting tomorrow's actions, then there should be no bias in today's decisions (see Laffont, 1980; Malinvaud, 1969; Simon, 1956).

This paper seeks to shed additional light on this issue. A basic point is that the stock nature of a pollution externality does not lead to irreversibilities unless learning is such that one might wish to emit negatively in the future to bring down pollution levels. In the terminology of Ulph and Ulph (1994b), one must be dealing with an *effective* irreversibility before there are any effects. This is also demonstrated in empirical work (Kolstad, 1993). If the level of uncertainty and the rate of learning are such that negative emissions are unlikely to be optimal in the future, then there is no irreversibility effect on the emissions side. Similarly, if the depreciation rate of the capital stock is sufficiently large relative to the rate of learning, such that negative investment in abatement capital is unlikely to be desirable, then there is no capital irreversibility. This does not mean that learning has no effect,[1] but that the irreversibility effect does not come into play.

In the next section of the paper, we set up a simple model of stock externalities and discuss known results within the context of the model, including the abatement and

---

[1] Ulph and Ulph (1994b) show that the fact that one is learning can change today's decisions, even without any sort of irreversibility effect.

environmental irreversibilities we have discussed. We then consider a simplified model illustrating our basic result. We close with conclusions.

2.2 A simple model of irreversibilities

The model we consider is a generalization of that of Freixas and Laffont (1984). Consider two time periods, with some action taken in each of these two periods. Thus for $t=1,2$, let U_t and x_t refer to the utility and action taken in time period t. Assume marginal utility is nonnegative and utility is twice differentiable and concave. Furthermore, assume utility in the second period depends on a random parameter, w, taken from a probability space Ω with a standard measure. After time period 2, some value w will be realized but from the perspective of either time period 1 or 2, there is uncertainty about what that value will be. Between time period 1 and 2, learning takes place and information is gained. We characterize what we have learned by a set S which is some partition of Ω; thus each element of S is a collection of elements of Ω. The set S contains elements which are themselves sets containing elements of Ω. In time period 1, we do not know which event in (element of) Ω is "correct;" i.e., will be ultimately revealed. Learning takes place between time periods 1 and 2. Before learning, we know the probability, π_s, that some $s \in S$ is "correct" (contains the true w); after the learning, we know which element of S is "correct" and thus have gained some information about what the true w, element of Ω, will be. We term another partition of Ω, S', a refinement of S if for all $s' \varepsilon S'$, there exists $s \in S$ such that $s' \subset s$. Thus a refinement will convey more information since it is a finer partition of Ω (see Laffont, 1989).

To solve or even specify this problem, we will work backwards from the last time period. Consider time period 2 first, given information $s \varepsilon S$ on Ω:

$$\max_{x_2} \mathscr{E}_{w \in s} [U_2(x_1, x_2, w)] \tag{2.1}$$

$$\text{s.t.} \qquad f_1(x_1) \leq x_2 \leq f_2(x_1) \tag{2.2}$$

If solutions to (2.1-2.2) exist, denote them by the set $X_2(x_1,s)$ which may or may not be a singleton and define $v(x_1,s)$ as

$$v(x_1, s) = \mathscr{E}_{w \in s}\,[U_2(x_1, X_2(x_1, s), w)]. \tag{2.3}$$

The problem in the first period is to find $x_1^*(S)$ that solves

$$\max_{x_1} \mathscr{E}_{s \in S}\,[U_1(x_1) + v(x_1, s)] \tag{2.4}$$

If a solution to (2.4) exists, our basic interest is in how x_1^* is affected by different information structures. Define $X_2^*(s) \equiv X_2(x_1^*,s)$. We can summarize current knowledge on the problem in four results.

The first result is due to Simon (1956). He shows that with U_i quadratic and no constraints (in eqn. 2.2), then the solution, x_1^*, is a function of the mean of w and no higher moments (certainty equivalence applies). This means that without constraint (2.2), there is no irreversibility effect in the quadratic problem and furthermore, there is no effect of risk aversion on the choice of x_1^*. This is of course due to the significant restrictions on the way in which uncertainty enters utility. Note that rather than restrict U_i and its domain so that a solution is guaranteed to exists, the theorem is confined in its applicability to cases where a solution does exist.

Theil (1957) and then Malinvaud (1969) relaxed these assumptions in developing the concept of first-order certainty equivalence. The functional form restrictions are relaxed at the expense of requiring a much less general representation of uncertainty. Malinvaud (1968) makes the assumption that the dependence of U_2 on w is small, that constraint (2.2) is non-binding everywhere, and that the Hessian of U_i is nonsingular in the vicinity of any optimal solution, $x_1^*,X_2^*(s)$. He then shows that x_1^* is independent of uncertainty (first-order certainty equivalence holds).[2]

Both of these results demonstrate that in some cases, uncertain parameters of a problem can be replaced with their expected values without distorting optimal decisions. Malinvaud's (1969) focusing on small amounts of uncertainty suggests that the result is not general; with significant amounts of uncertainty, different distributions of the random parameters may yield different first-period decisions. By extension, different rates of learning (i.e., refinements of S) may yield different x_1^*.

---

[2] This is a loose version of Malinvaud's result. He parameterizes uncertainty and compares the case of no uncertainty with small amounts of uncertainty and demonstrates that x_1^* is constant through a small neighborhood around the certainty point.

In both of these results, the constraints imposed by first-period actions on second-period actions had to be ignored (equation 2.2). This is demonstrated by a counterexample described in Laffont (1980). The constraints (2.2) characterize the irreversibility effect, first demonstrated by Henry (1974) and Arrow and Fisher (1974). Freixas and Laffont (1984) treat the case where equations (2.1-2.2) are of a very specific form:

Theorem 1 (Freixas and Laffont, 1984). In problem (2.1-2.4), assume f_1 and U_2 are independent of x_1 and $f_2(x_1) \equiv x_1$ so that eqn 2.2 becomes $x_2 \leq x_1$. If S' is a refinement of S, if $x_1^*(S')$ and $x_1^*(S)$ exist, and if v in eqn (2.3) is quasi-concave, then $x_1^*(S') \geq x_1^*(S)$.

Theorem 1 is a statement of the irreversibility effect: if today's actions restrict tomorrow's opportunities (equation 2.2), then more rapid rates of learning call for a bias in today's actions towards less restrictions. In environmental matters, if today's actions result in irreversible environmental damage[3], and one is acquiring information over time, then it is optimal to bias today's actions away from causing environmental damage.

In fact the Arrow and Fisher (1974) paper is a special case of this theorem. In that paper they discuss the problem of developing a unit of land vs. preserving the land. If x_i is the amount of land preserved in period i, the amount developed is $1-x_i$. The irreversibility is that once land is developed, it cannot be undeveloped. Thus in a two period world, $x_1 \geq x_2$. A direct application of theorem 3 is that when information is being acquired, x_1 should be larger (i.e., bias the decision in favour of the environment). This is the Arrow and Fisher (1974) result. A further implication is that the faster one is learning, the more land should be preserved in period 1. Thus the bias goes up with the rate of learning.

The most general result is due to Epstein (1980). Recall that for a partition S, consisting of elements s, the *ex ante* probability of a state s occuring is defined as π_s with π being the vector of these probabilities. Modify Eqn. 2.1 and 2.2 slightly to define $V(x_1,\pi,S)$ as

$$V(x_1, \pi, S) = \max_{x_2} \left\{ \sum_s \pi_s \mathscr{E}_{w \in s} \left[U_2(x_1, x_2, w) \right] \right\} \tag{2.5a}$$

---

[3] Irreversibility is not a biologic or physical term. Obviously, once you cut down a tree, you cannot re-root it, at least not immediately. However, if you are cutting down ten trees a year, cutting one more in one year is not irreversible since it can be corrected the following year.

$$\text{s.t.} \qquad f_1(x_1) \le x_2 \le f_2(x_1) \tag{2.5b}$$

Theorem 2 (Epstein, 1980). Consider two information structures, characterized by partitions S and S'. Suppose S is a refinement of S' and that in this context x_1^* solves (2.1-2.4) with partition S and x_1^{**} solves (2.1-2.4) with partition S'. If $V_x(x_1^*,\pi,S)$ (the derivative of V in eqn 2.5 with respect to x_1) is convex (concave) in π for all π in the simplex, then $x_1^* \le (\ge)\ x_1^{**}$.

While this theorem is quite general, it does not give very direct and transparent results. A number of authors have developed more definitive results for specific forms of the utility functions in eqn (2.1). Epstein (1980) himself presents several examples as do Hanemann (1989) and Ulph and Ulph (1994b).

Ulph and Ulph (1994b) provides the most general example; they develop results for a utility function of the form $U(x_1,x_2,w) = A(x_1) + B(x_2) + w\ C(x_1,x_2)$ with A, B and C quadratic. The basic result of Ulph and Ulph (1994b) is that in general learning has an effect on today's decisions, even without the presence of an irreversibility. This runs directly counter to the earlier results of Simon (1956) and Malinvaud (1968). However, these earlier results involved (more or less) quadratic approximations of utility. In the case of Ulph and Ulph (1994b), because w is multiplied by a quadratic, utility is not quadratic and thus the earlier results cannot be applied. The implication is that learning can have an effect, over and above that associated with irreversibilities. While their result is an important one, their functional form is still general enough that they do not always obtain definitive results when they apply their theorem to stock externalities (Ulph and Ulph, 1994a).

In our original problem, (eqn. 2.1-2.4), we were concerned with the trade-off between stock externalities and sunk abatement costs. Thus in the context of model (2.1-2.4), emissions are restricted from below as well as from above--emissions cannot go negative and they cannot go too high because of previous investments in abatement capital. Assume $\partial U_2/\partial x_1$ is independent of X_2. Before stating the next result, define two functions of information sets.

For any optimal solution to (2.1-2.4), $x_1^*(S)$, define A(S) and B(S) as

$$A(S) = \{s \in S | X_2(x_1^*, s) = f_1(x_1^*)\} \tag{2.6a}$$

$$B(S) = \{s \in S | X_2(x_1^*, s) = f_2(x_1^*)\} \tag{2.6b}$$

$A \subset S$ is the set of elements of S which end up resulting in actions at the lower bound of constraint (2.2). Similarly $B \subset S$ is associated with the upper bound of constraint (2.2). Without loss of generality, assume A and B are disjoint.

Theorem 3 (Kolstad, 1996). Assume a unique solution to (2.1-2.4) exists and $\frac{\partial U_2}{\partial x_1}$ is independent of X_2. Let $\tilde{x}_1$ be the first period output associated with uncertainty but no learning and assume (2.2) is not binding at $\tilde{x}_1$. Consider any information structure S resulting in first period optimal control $\hat{x}_1$. Then

$$sign \quad (\hat{x}_1 - \tilde{x}_1) = sign \quad [\Delta(S)] \tag{2.7}$$

where
$$\Delta(S) = \frac{\partial f_1}{\partial x_1} \mathscr{E}_{s \in A(S)} \mathscr{E}_{w \in s} \left[\frac{\partial U_2}{\partial x_2}\right] + \frac{\partial f_2}{\partial x_1} \mathscr{E}_{s \in B(S)} \mathscr{E}_{w \in s} \left[\frac{\partial U_2}{\partial x_2}\right] \tag{2.8}$$

For the proof of this, see Kolstad (1996).

This result can best be understood by focusing on equation (2.8). The first term on the right-hand side of equation (2.8) will typically be negative since it has to do with marginal utility at a lower bound of X_2. The fact that the lower bound is binding means that a reduction in X_2, were it possible, would increase utility. By a similar argument, the second term is positive. Thus whether $\Delta(S)$ in equation (2.7) is positive or negative depends on which term in equation (2.8) is stronger--the irreversibility associated with the lower bound or the one associated with the upper bound.

This effect can also be seen graphically. However, to see it most clearly let us simplify the problem somewhat. Suppose the event space consists of two possible states of the world: $\Omega = \{0,1\}$. If one knows $w=0$ then the lower bound of (2.2) is binding and if $w=1$ the upper bound is binding. If one only knows $w \in \{0,1\}$, then neither constraint in (2.2) is binding. Furthermore, assume U_2 is independent of x_1 and $0 < \partial f_1/\partial x_1 < \partial f_2/\partial x_2$.

Thus x_1 only influences x_2 through the constraints. These assumptions bascially simplify the graphical presentation.

Figure 2.1 shows the optimal choice of x_1, assuming no learning occurs before x_2 is chosen ($\tilde{x}_1$). However, if learning does occur, Figure 2.2 shows the optimal choice of X_2^* as a function of the state of the world, given $\tilde{x}_1$. Inspection suggests that equation (2.8) must be positive which implies $\hat{x}_1 > \tilde{x}_1$. This is the result of Theorem 3. Whichever extreme has the greater marginal utility/disutility will dominate and cause a bias in first period actions.

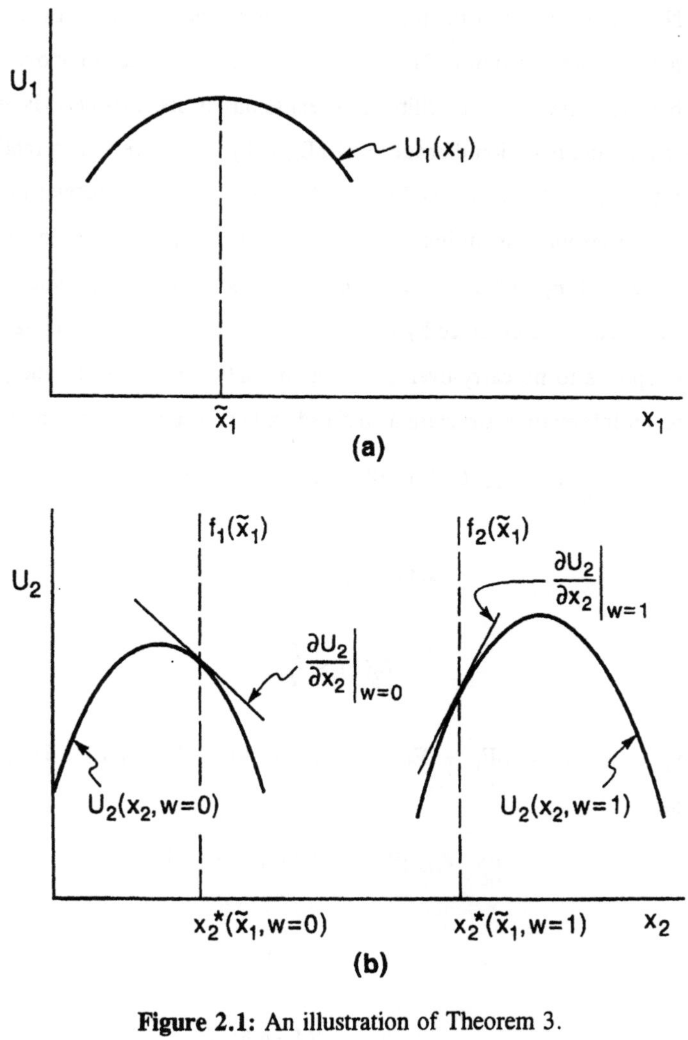

Figure 2.1: An illustration of Theorem 3.

2.3 Sunk costs and stock externalities

We now return to our original environmental problem, illustrating the applicability of Theorem 3. Consider a two period world in which a pollutant (E_i) is emitted in each period, $i=1,2$. Emissions increase utility on the one hand because they allow increased output of goods and services. On the other hand, emissions increase the stock of pollution which decreases utility. The stock of pollution cannot be reduced overnight through "negative emissions." Nor can the stock of pollution control equipment be reduced by uninvesting. In both cases, decay or depreciation is the only way to bring down the stock over time.

We thus write the positive utility from emissions in the two periods as $U(E_1)$ and $W(E_2)$ respectively and pollution damage as $D(\rho E_1 + E_2, w)$ where w is a random variable and ρ is the rate of persistence ($1-\rho$ is the rate of decay of emissions between periods 1 and 2). Thus $\rho=1$ corresponds to infinitely lived pollution and $\rho=0$ corresponds to no stock effect. Clearly E_1 and E_2 cannot be negative. The sunk nature of abatement capital with depreciation ($1-\delta$) can be represented by $E_2 \le E_1/\delta$. Thus $\delta=1$ means no capital depreciation and $\delta=0$ corresponds to no carry-over of pollution control capital from one period to the next. Let S be an information structure as defined earlier. The problem then is

$$\max_{E_1, E_2(s)} \mathscr{E}_{s\in S} \{U(E_1) + W(E_2(s)) - D(\rho E_1 + E_2(s), w)\} \tag{2.9a}$$

$$\text{s.t.} \quad E_1, E_2(s) \geq 0 \tag{2.9b}$$

$$E_2(s) \leq \frac{E_1}{\delta} \tag{2.9c}$$

If we let $x_1 = E_1$, $x_2 = \rho E_1 + E_2$, and $V(x_1,x_2,w) = W(x_2-\rho x_1) - D(x_2,w)$, we can rewrite this as:

$$\max_{x_1, x_2(s)} \mathscr{E}_{s\in S} \{U(x_1) + V[x_1, x_2(s), w]\} \tag{2.10a}$$

$$\text{s.t.} \quad x_1 \geq 0 \tag{2.10b}$$

$$\rho x_1 \leq x_2(s) \leq \frac{(1+\rho\,\delta)x_1}{\delta} \tag{2.10c}$$

The set A(S) defined earlier in the context of Theorem 3 corresponds to the situation where the lower bound in (2.10c) is binding. These are the events where pollution has turned out to be worse than expected and *ex post* it is desirable to emit a negative amount of pollution ($E_2 < 0 \Rightarrow x_2 < \rho x_1$) to bring down the stock of pollution. The set B(S) corresponds to the situation where the upper bound in (2.10c) is binding. These are the events where one wishes to undo the over-investment in abatement capital in period 1: pollution has turned out to be less of a problem and one wishes to uninvest in abatement capital.[4]

Let us assume the conditions of Theorem 3 are satisfied by U and V. The counterpart to equation (2.8) in this case is

$$\rho\, \mathscr{E}_{s \in A(S)} \frac{\partial V}{\partial x_2} + \frac{1 + \rho\,\delta}{\delta}\, \mathscr{E}_{s \in B(S)} \frac{\partial V}{\partial x_2} \qquad (2.11)$$

Typically, the first term of (2.11) is negative whereas the second term is positive. Consider first two extreme cases, one in which there is no stock effect of pollution ($\rho=0$) and the other where capital is perfectly fungible and there are no sunk costs ($\delta=0$).

In the first case ($\rho=0$), (2.11) is clearly non-negative, which from Theorem 3 means that any learning should result in larger (or no smaller) first period emissions. Since there is no pollution stock effect, one should be cautious in investing in too much control capital, which happens when emissions are controlled too much. In fact in our simple example, without any stock effect of pollution, there is no damage from first period emissions. Thus we would expect no difference.

It is a little more complex to examine the other case where there is no sunk control capital ($\delta=0$). Basically, the upper bound in equation (2.10) becomes nonoperational so B(S) is empty. Thus the second term in equation (2.11) is zero and consequently equation (2.11) is negative. From Theorem 3, this means that any learning should result in smaller first period emissions. In other words, when irreversibilities in capital do not exist, one should err on the side of the environment and under emit in the first period.

Since at the extremes we get either over- or under- emissions, clearly in the region between the extremes, the effect depends on the relative strength of ρ and δ. But it also depends on the magnitude of the expectations in equation (2.11). For instance if it is much

[4] While it is certainly true that abatement capital need not be used, if it is costless to do so and yields some benefits from pollution reduction, then it will be fully utilized.

more likely that one will eventually want to uninvest in control capital as opposed to negatively emit pollution, then the second term in equation (2.11) will dominate, leading to overall positivity of the expression. On the other hand, if learning is proceeding slowly enough compared with the decay rate of pollution and depreciation of control capital, the A(S) and B(S) may very well be empty and no bias will be called for.

We can examine this problem further by considering a numerical example. Consider the model expressed in equation (2.9) with utility logarithmic, of the form $U(E)=\ln(c_1+E)$ where c_1 is a constant >0, and damage quadratic, of the form $D(P,w)= c_w\ (c_2+P)^2$ where c_2 is a constant >0 and c_w is a state-dependent constant.[5] We use the following values of the various constants: $c_1=0.4$, $c_2=1$, $w=\{H,L\}$, $c_L=0.9$, $c_H=2.0$, prob$\{H\}=0.1$. We can then calculate the optimal level of first period emissions as a function of ρ and δ. Figure 2.2 shows the ratio of first period optimal emission with learning to first period emissions without learning, as a function of δ. The above discussion suggests that when there is no sunk control capital ($\delta=0$), the bias should be towards the environment and the ratio should be less than one (under-emit when learning occurs). This is shown in the figure. Furthermore, as δ moves towards 1, this bias reverses, and emissions are greater in the first period when learning occurs.

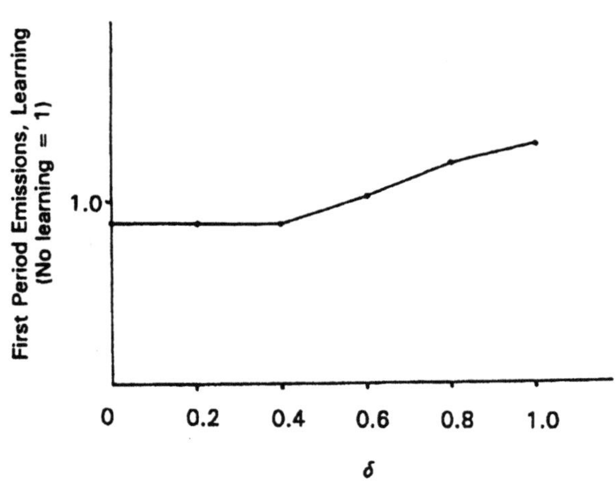

Figure 2.2: Ratio of first period emissions, learning to no-learning.

2.4 Conclusions

This paper has as its goal understanding the role of stock effects in pollution and sunk emission control capital on today's decisions on controlling pollution emissions. We have focused on the irreversibility effect as distinct from behavior induced by risk aversion. Using a simple two-period model we have demonstrated that only when it may be desirable to negatively emit or uninvest does one obtain a bias in today's control decisions due to irreversibilities. When today's actions only affect tomorrow's costs or utility but do not restrict tomorrow's choices, then there is no irreversibility effect. Of course, this result must be qualified by the specific assumptions of our Theorem 3.

The implications of this result for climate change policy are immediate. Unless the rate at which we are learning about climate change is such that we might wish to negatively emit in the future (to draw down the CO_2 stock), then there should be no tendency to under-emit today to forestall irreversible environmental effects.[6] Of course one may still wish to restrict emissions today to avoid low-probability catastrophic effects in the future. Such a bias in emissions control is due to risk aversion, not the irreversibility effect.

References

Arrow, K. (1968) Optimal Capital Policy and Irreversible Investment. In: J.N. Wolfe (Ed.) *Value, Capital and Growt*. Aldine, Chicago.

Arrow, K.J. and A.C. Fisher (1974) Environmental Preservation, Uncertainty and Irreversibility. *Quarterly J. Econ., 88, pp. 312-319.*

Epstein, L. G. (1980) Decision Making and the Temporal Resolution of Uncertainty. *Intl Economic Review, 21, pp. 269-83.*

Freixas, X. and J.J. Laffont (1984) *The Irreversibility Effect*. In: M. Boyer and R. Khilstrom (Eds.) *Bayesian Models in Economic Theory*, North-Holland, Amsterdam.

Gore, A. (1992) *Earth in the Balance*. Houghton-Mifflen, New York.

Kolstad, C. D. (1993) *Looking vs. Leaping: The Timing of CO_2 Control in the Face of Uncertainty and Learning*. In: Y. Kaya, N. Nakicenovic, W.D. Nordhaus and F.L. Toth (Eds.) *Costs, Impacts and Benefits of CO_2 Mitigation*, IIASA, Austria.

Hanemann, W. M. (1989) Information and the Concept of Option Value. *J. Env. Econ. and Mgmt., 16, pp. 23-37.*

Henry, C. (1974) Investment Decisions Under Uncertainty: The Irreversibility Effect. *Amer. Econ. Rev., 64, pp. 1006-12.*

[6] This result is illustrated in Kolstad (1993) using an empirical model of greenhouse gas control.

Kolstad, C. D. (1996) Fundamental Irreversibilities in Stock Externalities. *J. Public Economics*, forthcoming.

Laffont, J. J. (1980) *Essays in the Economics of Uncertainty*. Harvard University Press, Cambridge, Mass.

Laffont, J.J. *The Economics of Uncertainty and Information*. MIT Press, Cambridge, Mass.

Malinvaud, E. (1969) First Order Certainty Equivalence, *Econometrica, 37, pp. 706-718.*

Simon, H. (1956) Dynamic Programming Under Uncertainty with a Quadratic Criterion Function. *Econometrica, 24, pp. 74-81.*

Theil, H. (1957) A Note on Certainty Equivalence in Dynamic Planning. *Econometrica, 25, pp. 346-349.*

Ulph, A. and D. Ulph (1994a) *Global Warming: Why Irreversibility May Not Require Lower Current Emissions of Greenhouse Gases.* Discussion Paper 9402, Department of Economics, University of Southampton, UK.

Ulph, A. and D. Ulph (1994b) The Irreversibility Effect Revisited. *Mimeo.*

3 Who Gains from Learning about Global Warming?*

Alistair Ulph and David Ulph
University of Southampton and CSERGE
Highfield
Southampton, 5017 1BJ
United Kingdom

Abstract

In this paper we bring together two previously separate strands of literature dealing with global warming - the literature on the analysis of strategic interactions between independent national governments confronted with a dynamic problem of the the global commons, which has ignored issues of uncertainty and learning, and the literature on uncertainty, learning and irreversibility, which has assumed a single decision-making authority. The integration of these two literatures changes significantly the predictions derived from the separate literatures, especially the predictions about the impact of uncertainty and learning. Thus we show the following: (i) in situations where a single decision-maker would use the prospect of learning to delay cutting emissions, strategic interactions can cause countries to accelerate the cutting of emissions; (ii) while a single decision-maker is always better off when there is the possibility of learning, countries can now be worse off with the possibility of learning, and would not be prepared to pay anything for better information; this has the further implication that there is now an important additional source of gains from international agreements, or at least international coordination. There are two sources for these reversals of the conventional wisdom. One is that there may be states of the world where some countries have much higher damage costs than their expected values while others have much lower than expected damage costs; the strategic response in such states impose significant costs on all countries. The other

* Paper presented at conference on "The Economics of Global Environmental Change" University of Birmingham, May 9th -11th, 1994, and at workshop on "Designing Economic Policy for the Management of Natural Resources and the Environment", Crete, September 7th - 9th 1994. We are grateful to participants and to two anonymous referees for comments on earlier versions of this paper.

NATO ASI Series, Partnership Sub-Series, 2. Environment – Vol. 14
Economics of Atmospheric Pollution
Edited by Ekko C. van Ierland and Kazimierz Górka

source is significant asymmetries between countries, in particular cases where some countries face much higher uncertainty about potential damages than others. Both these sources of difficulty are plausible features of the global warming problem.

3.1 Introduction

It is well known that global warming has a number of features which make it a particularly difficult environmental problem to deal with: (i) it is a stock pollutant so that what is of concern is a long-term time path for emissions, with all the attendant difficulties of making long-term policy commitments; (ii) it is a global pollutant, in the sense that what causes damage is the global concentration of greenhouse gases to which emissions from all parts of the world contribute equally; this raises the familiar free-rider problem, which is compounded by the first problem by being a dynamic public good problem; (iii) there is considerable uncertainty about the both the scientific nature of the global warming problem and the possible costs and benefits of dealing with global warming but there is the possibility of learning more about these features in the future; (iv) there is no technology available for reducing already created concentrations of greenhouse gases, so greenhouse gas emissions are irreversible; combined with uncertainty and learning this raises the question of the timing of any policy to reduce emissions - should we delay cutting emissions of greenhouse gases until we learn that the costs of global warming are indeed serious or should irreversibility lead us to make deeper cuts in emissions now because if we do learn that damage costs are substantial we will be unable to undo the effects of current emissions?

The now voluminous literature on the economics of global warming touches on all these issues, but to the best of our knowledge there is no analysis which tries to combine all these features of the global warming problem. In this paper we present a preliminary version of just such an analysis. Of course it is a perfectly sensible research strategy to tackle a problem as complex as global warming in stages, so we need to say why we think it is important to bring the various features of the problem together. The particular questions we wish to focus on are the interactions between the first two features of the global warming problem, which we can think of as being the dynamic game aspects of the global warming problem, and the third and fourth features of global warming which we characterise as the irreversibility/learning aspects of global warming.

As far as the dynamic game aspects of global warming are concerned, there are two strands of the literature. Papers such as Hoel (1992a) and van der Ploeg and de Zeeuw (1992) use dynamic game theory to study the time paths of greenhouse gas emissions under both cooperative equilibrium, as would occur in a complete international agreement, and two non-cooperative equilibria - the open-loop Nash equilibrium, corresponding to a situation where individual governments act independently but can commit themselves to a path of future emissions, and the feedback Nash or subgame perfect Markov equilibrium where individual governments again act independently but now cannot commit themselves now to a future time path of emissions. As well as the rather obvious finding that global concentrations of CO2 will be higher under the non-cooperative equilibria than under cooperative equilibria, they derive the important finding that the feedback Nash equilibrium will lead to higher current emissions and higher final concentrations than the open-loop Nash equilibrium. The reason is that in the feedback equilibrium each government recognises that if they increase their current emissions then other governments will partially offset the effect of this by cutting back on future emissions; this provides all governments with an incentive to raise current emissions, which is not completely offset in the future. A second strand to the literature, of less relevance to this paper, concerns the question of how to design an international agreement to realise as much as possible of the gains to cooperation (see Barrett (1992), Maler (1991), Heal (1992) Hoel (1992b,c), Carraro and Siniscalco (1991) among many others).

The key point in terms of this paper is that all the analysis of the dynamic game aspects of global warming ignore the issues of uncertainty, learning and irreversibility. These issues are addressed in work by Kolstad (1992, 1993a,b,c), Beltratti, Chichilnisky and Heal (1993), Chichilnisky and Heal (1993), Manne and Richels (1993), Peck and Tiesberg (1993), Ulph and Ulph (1994a,b). As Ulph and Ulph (1994a) notes, the conventional wisdom is that irreversibility implies that if there is the possibility that in the future we will learn more about the potential costs and benefits of reducing emissions of greenhouse gases, then current emissions should be lower than lower than if there was no possibility of learning. However, this conventional wisdom is based on some simple models (Arrow and Fisher (1974), Henry (1974a,b), Freixas and Laffont (1982)) whose structure is inapplicable to global warming. It has been know since the work of Epstein (1980) that in a more general class of models the "irreversibility effect" can lead to current emissions with learning being higher or lower than current emissions without learning. In Ulph and Ulph (1994a, b) we showed that the sufficient conditions provided by Epstein for the irreversibility effect to take one sign or another are also inapplicable to the global warming problem. For the class of models we shall be dealing with in this paper what we showed in our (1994a) paper was that if we were to assume that greenhouse gases were reversible then the possibility of learning would in fact lead to *higher* current emissions than would be the case without learning; if greenhouse gases are irreversible, then a sufficient condition for current emissions with learning to be *lower* than current emissions without learning is that the irreversibility constraint should bite in the no learning case. As we argued in Ulph and Ulph (1994a) this is a very strong requirement; we are not aware of any of the empirical studies of optimal policy for greenhouse gas emissions in which this condition applies. So that, for the class of models we shall be using, there is a general *presumption* that, even allowing for irreversibility, current emissions will be higher when there is the possibility of learning than when there is no such possibility, unless the irreversibility constraint is very severe.

Again, the key point for this paper is that all the analysis of learning and irreversibility cited above assume a single decision maker, so would only apply if there was a single world government, or a complete international agreement. In short, the dynamic game analysis of the strategic interactions between governments ignores the issues of uncertainty, learning and irreversibility, while the analysis of uncertainty learning and irreversibility ignores the problem of strategic interaction between independent national governments. The purpose of this paper is to provide a very simple theoretical analysis of global warming which combines all these features. The questions we shall attempt to shed light on are the following.

(i) From the above discussion, it can be argued that there are two possible explanations of why we might expect governments not to take very much action to reduce current emissions of greenhouse gases: one is the "free rider" problem; the other is that they are waiting to see if global warming damage costs are really substantial. Do these two explanations reinforce each other or offset each other? We shall use our model to provide a more precise formulation of this question and show that, under some circumstances, the two arguments do not reinforce each other.

(ii) To understand the results in (i) we can pose a slightly different question. We have set out how the possibility of learning would affect the current emissions policy of a single world government. We would expect these results to apply to the cooperative equilibrium, and we shall show that this is indeed the case. But do these result carry over to non-cooperative equilibria? The answer is no. For example, we said that if greenhouse gas emissions were in fact reversible then, for our class of model, a single government would have higher current emissions with learning than without learning; this need not be true either for an individual country or for global emissions in non-cooperative equilibria. Thus non-cooperation can overturn the effects of learning that would apply to a single world government.

(iii) We know that, in a world of certainty, cooperation leads to lower emissions and concentrations of greenhouse gases and higher welfare than in non-cooperative equilibria. The same results apply to the case of uncertainty, both with and without learning. But the more interesting question is whether the gains from cooperation are higher or lower when there is learning or no learning; i.e. does the possibility of learning increase or decrease the incentives to reach an international agreement?

(iv) Finally, irrespective of how learning affects the current emissions policies of governments, for a single decision-maker the possibility of learning always makes the decision-maker better off than without learning. This leads to the question of how much a decision-maker would be willing to pay to acquire better information; Manne and Richels (1993), Peck and Tiesberg (1993) have carried out such calculations in the context of global warming. However, we shall show that this does not carry over to the case of non-cooperative behaviour with independent national governments; there are cases where either some governments, or even all governments, are worse off when there is the possibility of learning than when there is no learning.

We want to emphasise that the results described above are derived from a very simple theoretical model of global warming, for which we make no claims of realism. Our main message is that results derived from the dynamic game literature which ignores irreversibility and learning or from the literature on irreversibility and learning which ignores strategic interactions between governments, may not carry over to more general analyses which combine both features. We hope to explore these questions using more empirically based models of global warming, but this paper at least provides an agenda of questions to be posed. In the next section we shall set out our model. In section 3.3 we derive the results for the case where emissions of greenhouse gases are assumed to be reversible, since it is for this case that we can derive more precise analytical conclusions; in section 3.4 we extend the analysis to the case of irreversible emissions. Since it is not possible to derive all the results analytically, section 3.5 will present some numerical simulations, while section 3.6 offers conclusions and directions for future research.

3.2 The model

We capture the dynamic and strategic aspects of the global warming problem in the simplest way by assuming just two countries (indexed 1 and 2) and two time periods, indexed by 0 and 1. Each country has to choose its levels of emissions of greenhouse gases (treated as a single gas and measured, say, in terms of carbon equivalents) in each of the two periods. Damages from concentrations of greenhouse gases arise at the end of period 1; there is uncertainty about the extent of these damages for each country, which we capture by assuming that there are S states of the world, indexed s = 1, ..,S; the probability of state s occurring is denoted by $\pi_s \geq 0$, where $\sum_s \pi_s = 1$. We denote emissions for countries 1 and 2 in period 0 by e_0 and f_0 respectively; similarly emissions by countries 1and 2 in period 1 in state s are denoted by e_{1s} and f_{1s} respectively. The concentrations of greenhouse gases at the end of period 1 in state s are given by:

$$X_s \equiv \rho(e_0 + f_0) + e_{1s} + f_{1s} \tag{3.1}$$

where ρ represents a decay factor.

Countries 1 and 2 derive present value expected utilities from these emissions given respectively by U and V where:

$$U \equiv U_0(e_0) + \delta\{\sum_{s=1}^{s=S} \pi_s [U_1(e_{1s}) - a_s D(X_s)]\} \tag{3.2a}$$

$$V \equiv V_0(f_0) + \delta\{\sum_{s=1}^{s=S} \pi_s [V_1(f_{1s}) - b_s D(X_s)]\} \tag{3.2b}$$

U_0 and U_1 are respectively the benefits derived by country 1 from emissions in periods 0 and 1; a similar interpretation is given to V_0 and V_1 for country 2. Damages caused by the concentrations of greenhouse gases in each country are represented by two elements. First there is a damage cost function $D(.)$ which is common to both countries. Secondly, this common damage cost function is multiplied by a factor which can differ across states and countries as represented by the factors a_s and b_s for countries 1 and 2 respectively. a_S and b_S can vary across states in any way. We denote by $\bar{a}$ and $\bar{b}$ the expected values of a_s and b_s respectively. We assume that the utility functions are strictly concave and the damage function is strictly convex. The separability assumptions between benefits and damages from emissions, and the multiplicative form of the uncertainty parameters are clearly special.

We now turn to the information structure. There are two possible structures. Under *no learning* neither country will learn the true state of the world s until after they have chosen their period 1 levels of emissions; so in this case each country will have to set the same level of period 1 emissions for all states; under *learning* each country learns the true state of the world at the beginning of period 1, prior to choosing period emissions, and can thus condition period 1 emissions on the state of the world. Obviously this is the simplest possible learning structure we could adopt - learning is entirely passive, learning involves perfect information rather than just partial information (as in Kolstad (1992) for example) and learning occurs simultaneously for all countries. Again we are making no claims for realism, and are choosing the simplest model to keep the analysis tractable.

Closely related to the question of learning is the issue of irreversibility of greenhouse gas emissions, and here again we shall consider two cases. The *irreversible* case is where we assume that there is no possibility of undoing the effects of period 0 emissions in period 1 and we capture this simply by the constraints: $e_{1s} \geq 0, f_{1s} \geq 0$, for s = 1,...,S. In the *reversible* case we allow emissions in period 1 to be negative, so that period 0 emissions can be offset if required. This means that we must assume that the period 1 utility functions are defined over negative values, and the concavity assumption assures that countries will face increasing marginal costs of reversing period 0 emissions.

Finally we turn to the strategic structure of the model. We shall follow Hoel (1992a) van der Ploeg and de Zeeuw (1992) in considering three equilibria, one cooperative equilibrium and two non-cooperative. We model the *cooperative equilibrium* (or Pareto efficient equilibrium) by choosing emissions for both countries in both periods which maximise a social welfare function:

$$W \equiv \lambda U + (1-\lambda)V$$

where λ is a distributional weight that can take any value between 0 and 1. The first non-cooperative equilibrium we shall consider is the *open-loop Nash (OLN)* equilibrium, where each country takes as given the emissions of the other country in both periods and all states and chooses its own emissions policy to maximise its present value expected utility. There is a standard objection to the open-loop Nash equilibrium that while it is time-consistent (that is if both countries carry out the emissions levels in period 0 that are required by their OLN strategies then they will wish to carry out their period 1, state s, emissions as specified by their open-loop strategies) it is not *sub-game perfect*, in the sense that if either country deviates from its period 0 OLN emissions level then in general it will pay both countries to deviate from their OLN strategies in period 1; but this is not allowed for in the open-loop Nash equilibrium, so it is usually argued that this equilibrium concept would only be appropriate if the two countries could *commit* themselves in time period 0 to sticking to their period 1 OLN emissions levels irrespective of what actually happens in period 0. Since this seems an unlikely description of policy-making for global-warming we explore an alternative concept the *feedback Nash (sub-game perfect Markov) equilibrium (FBN)*. In this equilibrium, in period 1 each country can condition its emissions on the state of the game at the start of period 1; in this context the relevant state variable for the game at period 1 is the *stock* of emissions at the end of period 0: $X_0 \equiv \rho(e_0 + f_0)$. So each country will calculate a *policy* for how to set its emissions in period 1, state s, as a function of X_0, but taking as given the emissions of its rival in period 1 state s; we can thus solve for the Nash equilibrium in period 1 state s for any given value of X_0; in period 0 each country chooses its emission level taking as given the period 0 emissions of its rival and the policies of its rival for setting emissions in period 1. This feedback Nash equilibrium is a sub-game perfect Markov equilibrium in the sense that each country will wish to stick to its FBN policy for any given value of X_0. It is not fully sub-game perfect for that would require countries to condition their period 1 state s emissions on the *history* of the game up to that point; in the context of this simple model the only distinction between the two concepts is that conditioning on the history would allow policies to depend on the period 0 emission levels of the individual countries rather than just the stock; if there were more than two periods the distinction would have rather more force.

To summarise then we shall be considering twelve different cases - three equilibrium concepts for the cases of learning and no learning and reversible and irreversible emissions. It is partly because of the rather large number of cases that we need to compare that we have chosen to use such a simple model so that we can try to derive explicit solutions. In the next section we set out the solutions for the case of reversible emissions and in section 3.4 for irreversible emissions.

3.3 Reversible emissions

We begin with the case of no learning and then consider the modifications necessary for the learning case.

3.3.1 No Learning

In this case period 1 emissions cannot be conditioned on the state of the world, so that in the open-loop Nash equilibrium country 1 takes as given f_0, f_1 and chooses e_0, e_1 to maximise U, leading to first-order conditions:

$$U_0'(e_0) - \delta\rho\bar{a}D'[\rho(e_0 + f_0) + e_1 + f_1] = 0 \tag{3.3a}$$

$$U_1'(e_1) - \bar{a}D'[\rho(e_0 + f_0) + e_1 + f_1] = 0 \tag{3.3b}$$

These equations are just the usual condition that emissions in each period are chosen to equate marginal benefit with expected present value marginal damage cost. A similar set of conditions applies for country 2, giving four equations to solve for the four emission levels e_0, e_1, f_0, f_1. It is straightforward to check that the implicit reaction functions in (3.3a) and (3.3b) are downward sloping and stable.

For the feedback Nash equilibrium we note that (3.3a) and (3.3b) implicitly define e_1 and f_1 as functions of X_0 and thus gives the policy rules the two countries use for setting their period 1 emissions. Totally differentiating (3.3a) and (3.3b) yields:

$$-1 < \frac{df_1}{dX_0} = \frac{\bar{b}\,U_1''.D''}{(U_1''.V_1'' - \bar{a}D''V_1'' - \bar{b}D''U_1'')} < 0 \qquad (3.4)$$

with a similar expression for country 1. Any increase in the stock of emissions at the beginning of period 1 will lead to a reduction in emissions by each country in period 1, but not sufficient to completely offset the increase in the stock, so that the overall stock of emissions at the end of period 1 will increase. Similar results apply to country 2. For the feedback Nash equilibrium then we need to modify the first-order condition for period 0 to reflect the policy rules used by the players in period 1. Thus the first-order conditions for country 1 in the feedback Nash equilibrium are (3.3b) and :

$$U_0'(e_0) - \rho\delta\bar{a}D'[\rho(e_0+f_0)+e_1+f_1](1+\frac{df_1}{dX_0}) = 0 \qquad (3.3c)$$

If we compare (3.3a) and (3.3c) we see, from (3.4) that in the feedback Nash equilibrium the present value expected marginal damage costs of period 0 emissions are lower (by the factor in parenthesis that multiplies the second term in (3.3c)) than in the open-loop Nash equilibrium, so we would expect that in the feedback equilibrium both countries will produce higher emissions in period 0 than in the open-loop Nash equilibrium, and this in turn will lead to a higher level of final concentrations of greenhouse gases. This is just the result already derived by Hoel (1992a) and van der Ploeg and de Zeeuw (1992) and referred to in the previous section.

Finally we turn to the cooperative equilibrium, where e_0, e_1, f_0, f_1 are chosen to maximise social welfare. The first-order conditions for country 1's emissions are:

$$\lambda U_0'(e_0) - \delta\rho\psi D'[\rho(e_0+f_0)+e_1+f_1] = 0 \qquad (3.5a)$$

$$\lambda U_1'(e_1) - \psi D'[\rho(e_0+f_0)+e_1+f_1] = 0 \qquad (3.5b)$$

where:

$$\psi \equiv \lambda\bar{a} + (1-\lambda)\bar{b}.$$

A similar set of conditions applies for country 2. The interpretation of (3.5a) and (3.5b) is straightforward. The marginal gain in social welfare from emissions in a particular country in a particular period is equated to the present-value expected *global* marginal damage cost of those emissions. Thus, unlike the non-cooperative equilibrium where countries consider only the marginal damage of emissions to themselves, here it is the global level of marginal damages that is taken into account. Dividing (3.5a) and (3.5b) by λ and comparing with, (3.3a) and (3.3b) we see that marginal benefits of emissions in each period for country 1 are being equated to a larger level of marginal damages and so in general we will get the standard result of lower levels of emissions and concentrations in the cooperative equilibrium than in the non-cooperative equilibria.

3.3.2 Learning

If countries learn the true state of the world prior to their choice of emissions in period 1 then period 1 emissions can be conditioned on the true state of the world, s. To see how this modifies the results for the case of learning, we begin with the non-cooperative equilibria and note first that the first order condition (3.3b) for period 1 emissions by country 1 now becomes:

$$U_1'(e_{1s}) - a_s D'(X_0 + e_{1s} + f_{1s}) = 0, \qquad s = 1,\ldots,S \tag{3.6a}$$

where, again, (3.6a) implicitly defines the policy rule used by country 1 for setting its emissions in state s in period 1 as a function of X_0 and f_{1s}; the response of country 1's period 1 state s emissions to a unit increase in either the stock of emissions at the start of period 1 or the period 1 state s emissions of its rival will again lie between -1 and 0. It will be convenient to write the first-order condition for period 0 emissions in the non-cooperative equilibria more succinctly as:

$$U_0'(e_0) - \delta\rho \sum_s \{\pi_s a_s \phi_s D'(X_s)\} = 0 \tag{3.6b}$$

where in the open-loop equilibrium $\phi_s = 1$, while in the feedback Nash equilibrium, $\phi_s = 1 + \frac{df_{1s}}{dX_0}$ and $0 < \phi_s < 1$. For country 2 we define the corresponding term $\varphi_s = 1 + \frac{de_{1s}}{dX_0}$.

There will be a similar set of first-order conditions. The interpretation of these first-order conditions is exactly as before - marginal benefits of emissions are being equated to marginal present value expected damage costs; the difference is simply that in period 1 expected damage cost is now equal to actual damage cost in that state.

The cooperative equilibrium first-order conditions (3.5a) and (3.5b) get modified in a similar way to become:

$$\lambda U_0'(e_0) - \delta\rho \sum_s \{\pi_s \psi_s D'(X_s)\} = 0 \tag{3.7a}$$

and $$\lambda U_1'(e_{1s}) - \psi_s D'(X_s) = 0 \qquad (3.7b)$$

where $$\psi_s \equiv \lambda a_s + (1-\lambda) b_s$$

and a similar set of conditions apply for country 2's emissions.

To make further progress, and in particular to make comparisons between the equilibria with learning and those without learning it will be necessary to further specialise the model. As we showed in Ulph and Ulph (1994a), using the same model as the one we are using here, even in the case of a single country, comparison of emissions between the learning and no learning case is ambiguous. The reason is that the comparison depends on the concavity or convexity of a marginal value function, which in turn depends on the third derivatives of the utility and damage cost functions; the simplest set of assumptions to make, and which, in the single country case allowed us to make precise predictions, are that the third derivatives of theutility and damage cost functions are zero. Specifically, we assume:

$$U_t = -0.5(e_t * - e_t)^2 \qquad V_t = -0.5(f_t * - f_t)^2 \quad t = 0,1$$

$$D(X) = 0.5X^2$$

One way of interpreting this model is to recognise that in the absence of global warming country 1 would simply set emissions equal to e_t *, t = 0,1 and similarly for country 2; so we can think of e_t * as the emissions level that would arise in a Business-As-Usual (BAU) scenario, with quadratic damage costs and quadratic costs of reducing emissions below their BAU levels. These quadratic utility and damage cost functions make all the first order conditions linear, so it is a straightforward matter to solve the different equilibria. We can summarise the outcomes in the various equilibria in the following table 3.1.

Now a little manipulation of the formulae in table 3.1 yields the following results:

Proposition 1: In both the learning and no learning cases, there is the following ranking of aggregate period 0 emissions and end of period 1 concentrations of greenhouse gases:

$$(e_0+f_0)^{FN} > (e_0+f_0)^{ON} > (e_0+f_0)^{CN} \qquad (e_0+f_0)^{FL} > (e_0+f_0)^{OL} > (e_0+f_0)^{CL}$$

$$X^{FN} > X^{ON} > X^{CN} \qquad X^{FL} > X^{OL} > X^{CL}$$

where the superscripts F, O and C denote feedback Nash, open-loop Nash, and cooperative equilibria respectively, while the superscripts N and L denote no-learning and learning respectively.

Thus Proposition 1 essentially confirms for our model the results in terms of pollution emissions and concentrations found by Hoel (1992a) and van der Ploeg and de Zeeuw (1992); the new element is that their results were derived under conditions of certainty, while these allow for both uncertainty and learning. Nevertheless these results are not very surprising.

We now turn to the comparison of learning and no learning, and in particular the aggregate level of period 0 emissions. Now, in Ulph and Ulph (1994a), we showed that with for a single country, with the quadratic utility and damage cost functions used here, in the case of reversible emissions, period 0 emissions would be *higher* with learning than no learning, i.e. the possibility of learning would lead a single world government to postpone some of its reductions in emissions until it learns the true state of the world. Does this carry over to the case of many governments? Proposition 2 indicates what results are available.

Proposition 2: In the open-loop Nash and cooperative equilibria, there are higher aggregate period 0 emissions with learning than with no learning, i.e. $(e_0+f_0)^{OL}>(e_0+f_0)^{ON}$ and $(e_0+f_0)^{CL}>(e_0+f_0)^{CN}$.

Proof: From table 3.1, $(e_0+f_0)^{OL}>(e_0+f_0)^{ON} \Leftrightarrow \overline{H}<H$.

Define $c_s=a_s+b_s$ and $\eta(z)\equiv\dfrac{z}{1+z}$

Then, $\overline{H}=\sum_s \pi_s\eta(c_s)$, and $H=\eta(\sum_s \pi_s c_s)$, and the result follows from the concavity of the function η. A similar argument applies to the cooperative equilibrium. Q.E.D.

Now it is not very surprising that the cooperative equilibrium yields the same result as the case of a single government, since the cooperative equilibrium corresponds to a single world government maximising a social welfare function. It is rather more interesting that the result applies in the case of the non-cooperative equilibrium. Note two points. First, Proposition 2 tells us nothing about the feedback equilibrium, and as we shall see in section 3.5 the result need not be true for the feedback equilibrium. Second, Proposition 2 relates only to the aggregate of period 0 emissions; again as we shall see in section 3.5 the aggregate result need not apply to individual countries.

This completes the results that can be derived analytically for the case of reversible emissions. We now turn to the case of irreversible emissions.

Table 3.1

Solution variables	Non-Cooperative	Cooperative
No Learning		
X	$X^*/\{(1+\bar{a}+\bar{b})(1+\delta\rho^2 H)\}$	$X^*/\{(1+A)(1+\delta\rho^2 G)\}$
e_0+f_0	$\{e_0^*+f_0^*-\delta\rho H(e_1^*+f_1^*)\}/(1+\delta\rho^2 H)$	$\{e_0^*+f_0^*-\delta\rho G(e_1^*+f_1^*)\}/(1+\delta\rho^2 G)$
e_0	$e_0^*-\delta\rho\bar{a}\phi X$	$e_0^*-\delta\rho(1-\lambda)AX$
f_0	$f_0^*-\delta\rho\bar{b}\varphi X$	$f_0^*-\delta\rho\lambda AX$
e_1	$e_1^*-\bar{a}X$	$e_1^*-(1-\lambda)AX$
f_1	$f_1^*-\bar{b}X$	$f_1^*-\lambda AX$
Learning		
X_s	$X^*/\{(1+a_s+b_s)(1+\delta\rho^2\bar{H})\}$	$X^*/\{(1+A_s)(1+\delta\rho^2\bar{G})\}$
e_0+f_0	$\{e_0^*+f_0^*-\delta\rho\bar{H}(e_1^*+f_1^*)\}/(1+\delta\rho^2\bar{H})$	$\{e_0^*+f_0^*-\delta\rho\bar{G}(e_1^*+f_1^*)\}/(1+\delta\rho^2\bar{G})$
e_0	$e_0^*-\delta\rho\sum(\pi_s a_s\phi_s X_s)$	$e_0^*-(1-\lambda)\delta\rho\sum(\pi_s A_s X_s)$
f_0	$f_0^*-\delta\rho\sum(\pi_s b_s\varphi_s X_s)$	$f_0^*-\lambda\delta\rho\sum(\pi_s A_s X_s)$
e_{1s}	$e_1^*-a_s X_s$	$e_1^*-(1-\lambda)A_s X_s$
f_{1s}	$f_1^*-b_s X_s$	$f_1^*-\lambda A_s X_s$

where: $X^*=\rho(e_0^*+f_0^*)+e_1^*+f_1^*$

open-loop: $\phi=1$ $\qquad\varphi=1$

feedback: $\phi=(1+\bar{a})/(1+\bar{a}+\bar{b})$ $\qquad\varphi=(1+\bar{b})/(1+\bar{a}+\bar{b})$

$H=(\phi\bar{a}+\varphi\bar{b})/(1+\bar{a}+\bar{b})$

$A=\{\lambda\bar{a}+(1-\lambda)\bar{b}\}/\{\lambda(1-\lambda)\}$ $\qquad G=A/(1+A)$

open-loop: $\phi_s=1$ $\qquad\varphi_s=1$

feedback: $\phi_s=(1+a_s)/(1+a_s+b_s)$ $\qquad\varphi_s=(1+b_s)/(1+a_s+b_s)$

$H_s=(a_s\phi_s+b_s\varphi_s)/(1+a_s+b_s)$ $\qquad\bar{H}=\sum\pi_s H_s$

$A_s=\{\lambda a_s+(1-\lambda)b_s\}/\{\lambda(1-\lambda)\}$ $\qquad G_s=A_s/(1+A_s)$ $\qquad\bar{G}=\sum\pi_s G_s$

3.4 Irreversible emssions

When emissions are irreversible then, for the no learning case we need to add the pair of constraints: $e_1 \geq 0, f_1 \geq 0$, while for the learning case for each state s we add a similar pair of constraints $e_{1s} \geq 0, f_{1s} \geq 0$. Thus, for any of the three equilibrium concepts, in the no learning case there are four possible outcomes:

A $e_1 > 0, f_1 > 0$

B $e_1 = 0, f_1 > 0$

C $e_1 > 0, f_1 = 0$

D $e_1 = 0, f_1 = 0$

while for the learning case there are a similar set of four outcomes for each state s.

3.4.1 No Learning

In the no learning case, at least for the open-loop and the cooperative equilibria, it is possible to say when each of the four cases will arise. Thus:

Open-Loop Equilibrium.

Define $\hat{X} \equiv X^*/[1+(1+\delta\rho^2)(\bar{a}+\bar{b})]$; $X' \equiv \dfrac{\rho(e_0{}^*+f_0{}^*)+f^*{}_1}{1+\bar{b}+\delta\rho^2(\bar{a}+\bar{b})}$

$$X'' \equiv \frac{\rho(e_0{}^*+f_0{}^*)+e_1{}^*}{1+\bar{a}+\delta\rho^2(\bar{a}+\bar{b})}; \qquad \tilde{X} \equiv \frac{\rho(e_0{}^*+f_0{}^*)}{1+\delta\rho^2(\bar{a}+\bar{b})}$$

$$\hat{X} \equiv \frac{f_1{}^*}{\bar{b}} - \frac{\bar{a}f_1{}^* - \bar{b}e_1{}^*}{\bar{b}[1+(1+\delta\rho^2)(\bar{a}+\bar{b})]}; \breve{X} \equiv \frac{e_1{}^*}{\bar{a}} - \frac{\bar{b}e_1{}^* - \bar{a}f_1{}^*}{\bar{a}[1+(1+\delta\rho^2)(\bar{a}+\bar{b})]}$$

Then we have the following cases:

A If $\hat{X} < \min\{\dfrac{e_1{}^*}{\bar{a}}, \dfrac{f_1{}^*}{\bar{b}}\}$, then : $X = \hat{X}, e_1 = e_1{}^* - \bar{a}X, f_1 = f_1{}^* - \bar{b}X$

B If, $\dfrac{e_1{}^*}{\bar{a}} \leq \hat{X} < \hat{X}$, then: $X = X', e_1 = 0, f_1 = f_1{}^* - \bar{b}X$

C If, $\dfrac{f_1{}^*}{\bar{b}} \leq \hat{X} < \breve{X}$, then: $X = X'', e_1 = e_1{}^* - \bar{a}X, f_1 = 0$

D If, $\hat{X} \geq \max\{\frac{e_1 *}{\bar{a}}, \frac{f_1 *}{\bar{b}}\}$, then: $X = \tilde{X}, e_1 = 0, f_1 = 0$

In all cases $e_0 = e_0 * - \delta\rho\bar{a}X, f_0 = f_0 * - \delta\rho\bar{b}X$.

Of course Case A is just the same as the reversible case. Note that the conditions required for each case to occur involve disjoint sets of parameter space, but whose union covers all parameter values. This means that there always exists a unique open-loop Nash equilibrium.

Cooperative Equilibrium

Define: $\hat{X} \equiv \frac{X *}{[1 + (1 + A)(1 + \delta\rho^2)]}$ $\qquad X' \equiv \frac{\rho(e_0 * + f_0 *) + f_1 *}{[1 + (\lambda + \delta\rho^2)A]}$

$X'' \equiv \frac{\rho(e_0 * + f_0 *) + e_1 *}{[1 + (1 - \lambda + \delta\rho^2)A]}$ $\qquad \tilde{X} \equiv \frac{\rho(e_0 * + f_0 *)}{1 + \delta\rho^2 A}$

Then we have the following four cases:

A If $\hat{X} < \min\{\frac{e_1 *}{(1-\lambda)A}, \frac{f_1 *}{\lambda A}\}$, then: $X = \hat{X}, e_1 = e_1 * - (1-\lambda)AX, f_1 = f_1 * - \lambda AX$

B If $\frac{e_1 *}{(1-\lambda)A} \leq \hat{X} < \frac{f_1 *}{\lambda A}$, then: $X = X', e_1 = 0, f_1 = f_1 * - \lambda AX$

C If $\frac{f_1 *}{\lambda A} \leq \hat{X} < \frac{e_1 *}{(1-\lambda)A}$, then: $X = X'', e_1 = e_1 * - (1-\lambda)AX, f_1 = 0$

D If $\hat{X} \geq \max\{\frac{e_1 *}{(1-\lambda)A}, \frac{f_1 *}{\lambda A}\}$, then: $X = \tilde{X}, e_1 = 0, f_1 = 0$

For all four cases: $e_0 = e_0 * - \delta\rho(1-\lambda)AX, f_0 = f_0 * - \delta\rho\lambda AX$

Again it is clear that for any parameter values there must exist a unique cooperative equilibrium; this is scarcely surprising since we are simply solving a constrained optimisation problem.

However, for the feedback Nash equilibrium with no learning, while it is possible to write down the conditions under which each of the four cases must occur, as for the open-loop Nash and cooperative equilibria, the four sets of conditions do not in general span the entire range of parameter values, so that for some parameter values there may exist no feedback Nash equilibrium.

3.4.2 Learning

While in principle one could think of trying to do a similar exercise of identifying conditions on the underlying parameters when each possible equilibrium would arise, in practice such an exercise is infeasible, as there are 4^S possible configurations of equilibria (given that we have not so far put any restrictions on the parameters a_s, b_s). So we will need to rely on the numerical simulations reported in the next section for results on the case with learning.

For similar reasons it is not going to be possible to carry out the kind of analytical comparisons between the equilibria for different solution concepts and for the case of learning and no learning which we made in the case of reversible emissions, because we would need to be able to identify for any given solution concept which particular equilibrium (if any) would arise. So in the next section we present some numerical solutions of our model to help resolve some of the issues we have not been able to resolve analytically.

3.5 Numerical Results

To set up the numerical simulations we need first to specify the states of nature and the values of a_s, b_s for each state. As we have seen in the previous section the number of possible outcomes for equilibria in the case of learning with irreversible emissions is exponential in the number of states of the world, S, so it will be important to keep the number of states small. We have adopted the following simple structure. For each country we assume that the uncertainty parameter for damage costs can take two values: high or low; we denote by π_a and π_b the probabilities that countries 1 ans 2 respectively will face a high value for its damage costs, and by γ the correlation between the uncertainty parameters of the two countries. Then there are four possible states of the world and their structure and probabilities are given in table 3.2.

Table 3.2

State s	a_s	b_s	π_s
1	a_h	b_h	$\pi_a\pi_b + \gamma\omega$
2	a_h	b_l	$\pi_a(1-\pi_b) - \gamma\omega$
3	a_l	b_h	$(1-\pi_a)\pi_b - \gamma\omega$
4	a_l	b_l	$(1-\pi_a)(1-\pi_b) + \gamma\omega$

where $\omega = \sqrt{\pi_a\pi_b(1-\pi_a)(1-\pi_b)}$

It is clear from the last column of Table 3.2 that for given probabilities of high damages for countries 1 and 2 there are limits on the value of the correlation coefficient γ that will ensure that the probabilities of states 1,...,4 lie between 0 and 1.

We now turn to the simulation results, which are presented in Table 3.2. Since these contain a substantial number of results it will be sensible to explain fairly carefully the rationale for the numbers we have calculated. It will be useful to begin by saying what calculations we have carried out for each set of parameters that will help us to address the four questions posed in the Introduction. We take each question in turn.

(i) We said that there were two possible explanations why governments might take little immediate action to deal with global warming - the free rider problem and the waiting for better information problem; the question is whether these two explanations reinforce each other. To address this we shall focus on the total emissions of the two countries in period 0, which we denote by $t_0 \equiv e_0 + f_0$. The situation is shown in figure 3.1.

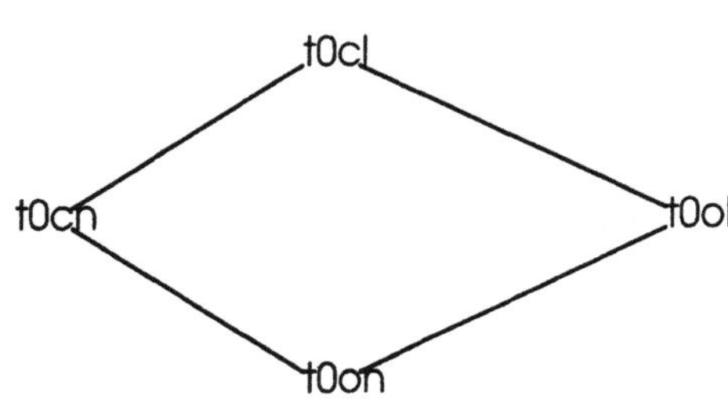

Figure 3.1

We take as our base the total emissions in period 0 in the cooperative equilibrium with no learning (*t0cn*).If we now introduce the possibility of learning we move to *t0cl*, and the argument is that we would expect that *t0cl* > *t0cn*. On the other hand if we move from the base by introducing non-cooperative behaviour, in this case open-loop Nash, then we move to *t0on*; and again, according to our argument, we would expect *that t0on* > *t0cn*. If we introduce both non-cooperative behaviour and learning we move to *t0ol*. We are interested in the ratio *t0ol/t0cn*, and how it decomposes into a factor due to non-cooperation and a factor due to learning. There are effectively two ways of doing this decomposition:

$$\frac{t0ol}{t0cn} = \left[\frac{t0ol}{t0cl}\right] \times \left[\frac{t0cl}{t0cn}\right] \tag{3.8a}$$

$$\text{or} \quad \frac{t0ol}{t0cn} = \left[\frac{t0ol}{t0on}\right] \times \left[\frac{t0on}{t0cn}\right] \tag{3.8b}$$

In the first decomposition the first factor represents non-cooperation, the second learning; in the second decomposition the first factor represents learning and the second non-cooperation. So we shall pose our first question by calculating the five ratios set out in (3.8) and asking whether in each decomposition the two ratios on the right hand-side are greater than 1, in which case our two explanations reinforce each other. Of course the same calculation can be done for the feedback solution, and we shall do these calculations for both the reversible and irreversible cases. Note that for the open-loop solution in the reversible case Propositions 1 and 2 have answered our questions, for they establish that the two factors on the right hand side of (3.8a) and (3.8b) are both greater than 1. So in this case learning and non-cooperative behaviour unambiguously reinforce each other to produce high current period emissions.

(ii) The second question we want to address is how the predictions about the effects of learning on current period emissions which we derived for a single government in Ulph and Ulph (1994a) carry over to the case of several governments. Recall that these predictions are (a) that in the case of reversible emissions we always get *higher* period 0 emissions with learning than no learning; and (b) that in the case of irreversible emissions, if the irreversibility constraint bites in the no learning case, then period 0 emissions will be *lower* with learning than no learning. We have seen in Proposition 2 that in terms of *total* emissions, t0, prediction (a) carries over to the open loop and cooperative equilibria. The simulations for answering question (i) will tell us whether, again for total emissions, predictions (a) and (b) carry over for other cases. But even if the predictions hold in aggregate, we can also ask whether predictions (a) and (b) also carry over for individual countries. Of course this question will only make sense if the countries are different in some respect. So to answer this part of question (ii) we shall calculate the ratio of period 0 emissions with learning to period 0 emissions without learning for each of the two countries and for the various combinations of solution concepts and reversible or irreversible emissions.

(iii) The third question we asked was how the possibility of learning would affect the gains from cooperation. To address this we shall measure the "gains from cooperation" by constructing the ratio of social welfare, W, in a non-cooperative equilibrium to that in a cooperative equilibrium; the lower the value of this ratio the greater the gains from cooperation. We shall compare this ratio for the case of learning and no learning, and we do this for four cases: open-loop or feedback concepts of non-cooperative equilibrium, and reversible or irreversible emissions.

(iv) Finally we asked whether in a model of several governments interacting strategically it was the case that countries would always be better off with the possibility of learning than without such a possibility, as must be the case for a single decision-maker. To assess this we calculate the ratio of expected present value utility for each country (i.e. U and V for countries 1 and 2 respectively) with learning to that without learning; we do this for different equilibrium concepts and for reversible and irreversible emissions.

The simulation results are presented in table 3.3 (presented in the appendix). We have used 16 different sets of parameters, and for each set we present the results for the case of reversible and irreversible emissions. There are 14 parameters in this model, and these are presented in the first 7 rows of table 3.3, with the first 7 in the REV column, the second 7 in the IRREV column. The parameters are respectively: the probabilities of high damages for countries 1 and 2 and the correlation coefficient between the two damage parameters; rather than setting the values for high and low damages it is more instructive to set the mean and variance of the damage parameters and these are the next four parameters; the next four parameters are the Business as Usual (BAU) values of emissions in each period for each country; finally there is the decay factor, the discount factor and the share of country 1 in the social welfare function. With the large number of parameters we cannot claim to have conducted an exhaustive search over parameter space. In the first 10 sets we assume that the two countries are identical, and investigate the effect of changing the main parameters systematically. Cases 1-3 investigate the effect of increasing the variance of uncertainty about the damage parameters; cases 4-6 do the same, but for a higher level of mean damages; cases 3 and 4 have the same absolute level of variance; cases 3 and 5 have the same ratio of variance to mean; we take the parameters of case 5 as a "base case" and investigate changes in other parameters. All the cases so far have assumed zero correlation between the damages of the two countries; cases 7 and 8 investigate the effect of having correlation coefficients of 1 and -1.

Finally, all the cases so far assume that BAU levels of emissions are the same in the two periods; case 9 investigates the effect of BAU emissions in period 1 being lower than period 0, corresponding say to rapid improvements in energy saving technology relative to economic growth; case 10 investigates the opposite case where BAU emissions rise over time. The final six cases investigate the effect of introducing asymmetries between the two countries. Case 11 gives the two countries different variances of uncertainty (interpreted as different gains from learning); case 12 additionally gives the country with high variance a higher probability of high damages. Case 13 looks at a different form of asymmetry - one country has a higher level of BAU emissions than the other and hence higher marginal benefits from emissions; cases 14 and 15 combines the two previous forms of asymmetry with the country having higher BAU emissions having respectively the higher and then the lower variance of uncertainty about damages. Finally case 16 is the same as 15 but with correlation coefficient of 1 instead of 0.

Turning to the rows of Table 3.3, these present the various ratios we have discussed above. Rows 8 to 17 present the ratios of total period 0 emissions designed to answer the first question; the first five are for the open-loop Nash equilibrium, the second five for the feedback Nash equilibrium; thus t0olcn is the ratio of total period 0 emissions in the open-loop equilibrium with no learning to those in a cooperative equilibrium with no learning. Rows 18 to 35 present the ratios designed to answer questions 2 and 4, dealing with comparisons between what happens with learning and without learning; rows 18 to 23 are concerned with the open-loop equilibrium and construct ratios of, respectively, period 0 emissions for countries 1 and 2, final concentrations of pollution, expected utility for countries 1 and 2 and social welfare for the cases of learning to those of no learning; rows 24 to 29 present the same information for the feedback equilibrium and rows 30 to 35 for the cooperative equilibrium. The final four rows present the ratios of social welfare in noncooperative equilibrium to those in cooperative equilibrium (our measure of "gains to cooperation"); these are done for the two concepts of non-cooperative equilibrium and the cases of no learning and learning.

We now assess what light our simulation results through on the four questions we have posed.

(i) By looking at rows 8 and 13 we see that, for both concepts of a non-cooperative equilibrium, period 0 emissions in a non-cooperative equilibrium with learning are always higher than in a cooperative equilibrium with no learning. We want to ask what is the contribution of the two possible explanations of "lack of cooperation" and "waiting to learn" to explaining this increase in current period emissions. We saw above that there were two different ways of decomposing the effect on period 0 emissions of moving from the cooperative equilibrium with no learning to a non-cooperative equilibrium with learning into an effect due to non-cooperation and an effect due to learning. Table 3.4 presents these two decompositions into the two factors; we want to know whether these factors are positive or negative, i.e. whether the particular ratios are greater than 1 or not. We examine this for the four cases corresponding to different non-cooperative solution is used and whether or not emissions are reversible. Table 3.4 indicates whether the appropriate ratio is always greater than 1 (A), sometimes greater than 1 (S) or never greater than 1 (N).

Table 3.4

	Decomposition1	*Decomposition1*	*Decomposition2*	*Decomposition2*
	Cooperation	Learning	Cooperation	Learning
	t0olcl	t0clcn	t0oncn	t0olon
O-L REV	A	A	A	A
O-L IRREV	A	S	A	S
FB REV	A	A	A	S
FB IRREV	A	S	A	N

Recall that the first row of table 3.4 does not depend on simulation results - these follow from Propositions 1 and 2. Thus we see from Table 3.4 that lack of cooperation always leads to higher period 0 emissions (less action by governments to cut emissions) irrespective of solution concept or whether or not emissions are reversible. The learning factor however can lead to lower first period emissions. In the first decomposition, the learning factor applies to cooperative equilibrium, and here learning only leads to lower period 0 emissions and that is case 9 when there are irreversible emissions. We shall discuss this in more detail below with respect to the next question. In the second decomposition the learning factor applies to non-cooperative equilibrium, and here we see a much stronger tendency for learning to lead to lower period 0 emissions than without learning; as our theory predicts, with open-loop this can only happen with irreversible emissions; with feedback equilibrium and irreversible emissions, for the simulations we report we always have lower period 0 emissions with learning; however other simulations, not reported here, show that this is not a general result.

(ii) In answering our second question we explore more carefully the circumstances under which learning leads to higher or lower period 0 emissions. Recall that for a single decision-maker with reversible emissions we would always get higher period 0 emissions with learning, while with irreversible emissions a sufficient condition to get lower period 0 emissions with learning would be that the irreversibility constraint bites in the no learning case. We would expect these results for the single decision-maker to apply to the cooperative case. In terms of total emissions it is certainly the case that when emissions are reversible we get higher period 0 emissions with learning than without learning in the cooperative equilibrium. The only case where we get lower period 0 emissions is in case 9 when emissions are irreversible; it turns out that this is indeed a case where the irreversibility constraint bites in the no learning case. The reason why this appears to happen in case 9 is that this is the case where BAU emissions are lower in period 1 than period 0, and the irreversibility constraint is more likely to bite the lower is the demand for emissions in period 1. So the cooperative equilibrium appears to follow the predictions derived in Ulph and Ulph (1994a) for a single decision-maker)

Turning to the non-cooperative equilibria, we saw in the previous section that, in terms of total emissions, we were much more likely to get lower period 0 emissions with learning than was the case with cooperative equilibrium. Note first that for reversible emissions, while open-loop we know always gives higher period 0 emissions with learning than no learning, this is not the case for the feedback equilibrium , where, apart from cases 3 and 7, period 0 emissions are lower with learning than no learning. With irreversible emissions, in both the open-loop and feedback cases we get lower period 0 emissions with learning than no learning;

indeed, as already noted this always happens for the feedback results reported here. Note that these can occur even although the irreversibility constraint does not bite in the no learning (e.g. in case 9). Of course, having the irreversibility constraint bite in the no learning case was only a sufficient, not a necessary condition, for lower period 0 emissions with learning in the case of a single decision-maker. In general then, the introduction of non-cooperative behaviour violates the predictions we get from the case of a single-decision-maker; in general we get more occurrences of learning lowering period 0 emissions than we would predict from that analysis. We shall give an intuition into why this should occur when we consider the next question.

All the above discussion has been in the context of total period 0 emissions by the two governments; obviously this carries over to individual governments for the symmetric case of identical countries. What can be said about the cases where countries are not identical? The interesting result is illustrated in case 11, where country 2 has a higher variance than country 1 (and hence more to gain from learning). In the case of reversible emissions, although total period 0 emissions are higher with learning than without learning, this is true only for country 2. This is not surprising; theory predicts this would be true of a single-decision-maker and of the two countries in total, so it is not surprising that this learning effect will apply to the country which has most to gain from learning; but the high period 0 emissions of country 2 cause country 1 to cut back its emissions in the learning case. So country 2's response to learning imposes an externality on country 1, and not surprisingly country 2 is better off with learning than without while the reverse is true for country 1. We shall return to this. The prediction that the country with higher variance will increase it period 0 emissions with learning while the one with lower variance cuts its period 0 emissions carries over to cases 12, 14, 15, and 16; it is true whether emissions are reversible or irreversible, and it is also true when the feedback equilibrium is used. The other asymmetry we have examined (differences in BAU emissions) appears not to affect differentially the impact of learning on period 0 emissions of the two countries (see case 13).

(iii) We now address the question of whether the gains from cooperation are greater with learning or without learning, where the gains to cooperation are measured by the ratio of social welfare in the non-cooperative equilibrium to social welfare in the cooperative equilibrium; as we can see from the last four rows of table 3.3 this ratio is always less than 1, as we would expect, indicating that there are always to be made from cooperation. With the exception of case 7, this ratio is always lower with learning than without learning, indicating that, with the exception of case 7, there are greater gains to be had from cooperation when learning is taking place. To see why this result arises, and why case 7 provides the exception, it is useful to compare cases 7 and 8 which differ only in the correlation coefficient between damages in the two countries; in case 7 damages are perfectly correlated; in case 8 they are perfectly negatively correlated. Table 3.5 presents the emissions chosen by the two countries for the open-loop and cooperative equilibria with learning and no learning, reversibility and irreversibility, because of the symmetry of these cases we only present the policies for country 1.

In case 7 the only two states that can occur are 1 (both have high damages) and 4 (both have low damages); in case 8 the only two states that can occur are 2 (country 1 high damages, country 2 low damages) and state 3 (country 1 low damages, country 2 high damages).

Let us begin with the no learning cases. Note first that in both open-loop and cooperative equilibria we get the same outcome in cases 7 and 8; this is because the difference in correlation coefficient makes no different to expected damages and the countries are going to have the same level of emissions across all states. Moreover, since the level of emissions is positive in period 1, the irreversibility constraint does not bite. So the only difference is between cooperative and non-cooperative equilibrium, and this has the usual relationship that there are higher emissions in both periods in the non-cooperative equilibrium, and both countries are worse off in the non-cooperative equilibrium, yielding the standard gains to cooperation.

Now turn to the learning case, and begin with the cooperative equilibrium. In case 7, if both countries turn out to face high damages they will want to cut their emissions substantially (setting negative emissions in the reversible case) and if they turn out to have low damages they will raise their emissions; because they are fine tuning their emissions in period 1 to the state of nature, they will raise their emissions in period 0, just as a single decision-maker would do, (see Ulph and Ulph (1994a) for more explanation), and this fine tuning means that both countries are clearly better off with learning than without learning. If emissions are irreversible, they cannot have negative emissions in state 1, so to compensate they set somewhat lower emissions in period 0 (but still higher than in the no learning case), and this allows them to set somewhat higher emissions in state 4. Case 7 then delivers all the results derived for a single decision-maker - higher period 0 emissions and welfare in learning than no learning. Case 8 is more interesting. In a cooperative equilibrium, if countries are identical in

Table 3.5

	CASE 7	CASE 7	CASE 7	CASE 7	CASE 8	CASE 8	CASE 8	CASE 8
VAR.	π_s	no learn	learn	learn	π_s	no learn	learn	learn
			rev	irrev			rev	irrev
		open loop	open loop	open loop		open loop	open loop	open loop
e0	n.a.	26.17	32.24	32.24	n.a.	26.19	26.19	18.55
e11	0.5	17.07	0.14	0.14	0.0	n.a.	n.a.	n.a.
e12	0.0	n.a.	n.a.	n.a.	0.5	17.07	-28.89	0.0
e13	0.0	n.a.	n.a.	n.a.	0.5	17.07	63.03	60.61
e14	0.5	17.07	48.14	48.14	0.0	n.a.	n.a.	n.a.
U	n.a.	3.08	3.17	3.17	n.a.	3.08	2.92	2.93
		coop	coop	coop		coop	coop	coop
e0	n.a.	17.83	23.67	19.67	n.a.	17.83	17.83	17.83
e11	0.5	7.29	-5.93	0.0	0.0	n.a.	n.a.	n.a.
e12	0.0	n.a.	n.a.	n.a.	0.5	7.29	7.29	7.29
e13	0.0	n.a.	n.a.	n.a.	0.5	7.29	7.29	7.29
e14	0.5	7.29	34.16	35.86	0.0	n.a.	n.a.	n.a.
U	n.a.	3.16	3.24	3.23	n.a.	3.16	3.16	3.16

terms of their BAU emissions, (and hence have the same marginal abatement costs), they will set the same level of emissions as each other no matter what the state, since they essentially set marginal abatement costs equal to the sum of the marginal damage costs across the two countries. But in states 3 and 4 the sum of marginal damages is the same; the only difference is who has the high and low damages, but that is irrelevant for a cooperative equilibrium. So in case 8 there must be the same level of emissions by each country in each state that occurs with positive probability; but if there is no difference in behaviour across states then learning is irrelevant and we must have the same equilibrium in the case of learning as no learning, as indeed is the case; moreover since period 1 emissions are positive irreversibility is irrelevant. So now there is no gain from learning in a cooperative equilibrium when damages are perfectly correlated and countries are identical; in essence there is no social uncertainty in this case.

Now turn to the open-loop equilibrium. In case 7, matters are rather similar to the cooperative equilibrium. When both countries have high damages both cut their emission levels; when both have low damages they both raise their emission levels (relative to the no learning equilibrium). For exactly the same reasons as in the cooperative equilibrium both countries raise their period 0 emissions relative to the no learning case, and both countries end up better off than in the no learning case; since emissions are positive in states 1 and 2 irreversibility is irrelevant. Of course, as in the no learning case, the level of emissions in the non-cooperative equilibrium are higher in all periods and all states than in the cooperative equilibrium, so there are the usual gains from cooperation, but the gains from cooperation are (slightly) lower in the case of learning than no learning, and this we interpret as essentially a kind of diminishing returns to cooperation - since learning in this case is delivering higher utility to each country than in the no learning case, there is slightly less to be gained from cooperation.

The situation is strikingly different in case 8. In state 2, country 1 faced with high damages cuts its period 1 emissions, but country 2 faced with low damages raises its emissions; but they do not cut or raise them to the amounts suggested in case 7; they substantially overshoot. The reason is clear. Country 1 is cutting its emissions because it believes country 2's emissions will stay the same and it is trying to cut the overall level of emissions because of its high damages; but country 2 is following the same logic but in reverse, and is raising its emissions, so country 1 has to make even deeper cuts in emissions. In the reversible learning case we actually end up with *exactly the same level of aggregate emissions as in the no learning case, and, since this leads to the same level of period 0 emissions, with the same level of concentrations as in the no learning case.* But this is a very much worse outcome for the two countries than in the no learning case, for two reasons; first, there is now a substantial inefficiency in achieving the same target of total emissions as in the no learning case, for country 1 has a much higher marginal abatement cost than country 2; second, the country with high damage costs now has to face an additional cost - the extra cost on unnecessarily having to cut back on its emissions; of course country 2 is very much better off in state 2, because as well as low damages it gets the benefit of much higher emissions. These positions are exactly reversed in state 3, but what this means is that, compared to the no learning equilibrium, each country now faces considerably greater variation in its *ex post* utility levels than would be the case if they just kept to the no learning policy, and since they are risk averse this is a further factor reducing *ex ante* expected utility. So the two countries are worse off with learning than no learning; since the cooperative equilibrium gives the same level of expected utility with learning as with no learning there are significantly greater gains to be had from cooperation with learning than with no learning. The source of these extra gains from cooperation are now clear: they are the prevention of the changes in emissions that arise in states where one country has low damage costs and the other has high damage costs, where one country is trying to raise the overall level of emissions and the other is trying to lower it; both countries cannot succeed and all they are doing in the process is creating inefficiencies in abatement and greater variability in utility. One final interesting point emerges if we look at the irreversible case in the non-cooperative equilibrium in case 8. Because emissions cannot go negative this puts a brake on the fruitless competition between the two countries to change aggregate emissions in different directions, and actually means that the two countries are better off with irreversible emissions than with reversible emissions. This is in contrast to what happens with a single decision maker or in a cooperative equilibrium (see case 7) where irreversibility is a constraint on what agents can do and so lowers expected utility (if the constraint bites).

Now the two cases we have examined in some detail are extreme cases of correlation, in which only two of the four possible states have positive probability; all the other simulations we have considered assume zero correlation, and will give positive probability, in general, to all four states. These more general cases will thus contain a mix of the two kinds of considerations that arise in cases 7 and 8. As we have seen, in all these cases of zero correlation we also obtain the result that there are greater gains from cooperation with learning than no learning, which suggests that the factor that arose in case 8 dominate that which arose in case 7; this is not surprising since the lower gain in cooperation under learning that arose in case 7 was rather slight while the greater gain in cooperation that arose in case 8 was quite substantial. So this suggests that as long as there is a reasonable probability attached to states 2 and 3 occurring we can expect greater gains from cooperation when there is the possibility of learning.

The explanation given above of why we get greater gains from cooperation with learning also explains why we are more likely to get lower period 0 emissions with learning than we would have predicted from the analysis of a single decision-maker, because an obvious way of reducing the additional costs imposed on countries in states 2 and 3 caused by the interaction between strategic behaviour and learning is to cut emissions in period 0.

(iv) We can now deal more briefly with the final question, because our analysis of the previous questions has already provided much of the answer. Now note that in the case of a single decision-maker, whatever happens to period 0 emissions and the like, the decision-maker is always better off with learning than without, for the simple reason that the decision-maker can always carry out the same policy with learning as without learning, and so if another policy is chosen that can only be because it makes the decision-maker better off. We assess how far this is modified by introducing strategic behaviour by our two countries.

For the symmetric cases, we have seen that in the open-lop equilibrium countries are always worse off with learning than no learning except for case 7; the reason for this has already been given - when countries learn that the true state is 2 or 3 their strategic response in these cases imposes significant costs on both countries, and if these states have a reasonable probability of occurring these costs outweigh the more usual source of benefits from learning that arise in states 1 and 4. For the cooperative equilibrium, countries are always better off with learning than with no learning, which is just a direct analogue of the single decision-maker case.

For the asymmetric cases, as we have already noted, the country with a lower variance than the other ends up being worse off with learning; the reason is that the country with the higher variance raises its period 0 emissions compared to the no learning case by more than the other country would like; so the other country has to cut its period 0 emissions, and this imposes more costs on the low variance country. So again the strategic interaction between the two countries means that one country's response to learning imposes an externality on the other country.

Thus for a wide range of cases, either one or both country is worse off with learning than no learning; clearly these countries would not be prepared to pay anything for the possibility of getting better information.

This completes our use of the simulation results to address the four questions we posed in the Introduction. We summarise our main conclusions in the next section.

3.6 Conclusions

In this paper we have brought together two different strands of literature in relation to global warming - the analysis of strategic interactions between independent national governments choosing their time paths for emissions of greenhouse gases, which has ignored issues of uncertainty and learning, and the analysis of uncertainty, irreversibility and learning which has assumed a single decision-maker and so ignored the issues of strategic interaction between governments. This paper has sought to explore how the analysis of a model which combines all these features might change the conclusions derived from the two separate literatures. We have shown that there are important new insights to be derived from bringing these features together. The main difference is that strategic interaction alters the conclusions about the impact of learning derived from the literature on a single decision-maker: learning is much more likely to cause current emissions to be *lower* than they would be without learning, and this can be true of individual countries or of all countries; in other words, the fact that it may be possible to learn more in the future about the extent of damages is much less of a reason for governments to delay reducing emissions than would be the case if there was a single world government. Moreover, learning can now make countries worse off than they would be without learning, quite contrary to the situation with a single decision-maker, so that countries would not be prepared to pay anything to gain better information; on the other hand this can mean that there are significantly greater gains to be had from cooperation when there is the possibility of learning.

There are two principle factors contributing to these results. First, even if countries are completely identical, if there is not perfect correlation between the uncertain damages faced by different countries from global warming, which seems to be a realistic description of the problem, then there will be states of the world with positive probability where some countries face above average damages and others face below average damages; strategic interactions will mean that high damage countries will be cutting their emissions to try to cut aggregate emissions while low damage countries will be raising their emissions in order to try to raise aggregate emissions. These moves are largely self-defeating but impose substantial costs of inefficiencies in abatement of emissions and increase the variability of individual countries' welfare, lowering expected utility. Countries would be better off agreeing to stop such actions, and this is an additional source of gains from cooperation over and above the usual gains to be had from a collective lowering of emissions; to distinguish the two, perhaps the former gains are better referred to as gains from coordination rather than cooperation. Second, differences between countries can cause different predictions about the effect of learning; the difference that we have found to be important is where countries face differences in the variance of their uncertain damages; countries with high variance respond to learning by raising their current period emissions (i.e. adopting a wait-and-see strategy), but this causes countries with a low variance to have to cut their current emissions; the low variance countries are worse off as a result of learning, the high variance countries better off.

Of course these results have been derived in an extremely simple theoretical model, but one which we believe is rich enough to capture the relevant features of the problem and to deliver interesting results. Even within this simple framework, we have had to resort to some numerical calculations, and the simulations we have carried out have been far from exhaustive. So there is clearly much more to do in exploring this model and extending it in obvious ways. More importantly though, we would like to try to explore these ideas using models which are designed to capture the real-world features of global warming and actual countries, and we hope to be able to report results of that in a future paper.

Appendix

Table 3.3a

pars/vars	case 1	case 1	case 2	case 2	case 3	case 3	case 4	case 4
	REV	IRREV	REV	IRREV	REV	IRREV	REV	IRREV
πa e0*	0.5	80.0	0.5	80.0	0.5	80.0	0.5	80.0
πb e1*	0.5	80.0	0.5	80.0	0.5	80.0	0.5	80.0
γ f0*	0.0	80.0	0.0	80.0	0.0	80.0	0.0	80.0
$\bar{a}$ f1*	0.5	80.0	0.5	80.0	0.5	80.0	0.75	80.0
va ρ	0.1	0.95	0.15	0.95	0.2	0.95	0.2	0.95
$\bar{b}$ δ	0.5	0.9	0.5	0.9	0.5	0.9	0.75	0.9
vb λ	0.1	0.5	0.15	0.5	0.2	0.5	0.2	0.5
t0olcn	1.545	1.481	1.596	1.478	1.656	1.484	1.570	1.389
t0olcl	1.414	1.356	1.379	1.306	1.332	1.268	1.438	1.315
t0clcn	1.092	1.092	1.157	1.132	1.243	1.171	1.092	1.056
t0olon	1.058	1.015	1.094	1.013	1.134	1.017	1.069	0.946
t0oncn	1.460	1.460	1.460	1.460	1.460	1.460	1.469	1.469
t0flcn	1.858	1.753	1.865	1.717	1.883	1.684	2.091	1.859
t0flcl	1.701	1.605	1.611	1.517	1.515	1.439	1.916	1.760
t0clcn	1.092	1.092	1.157	1.132	1.243	1.171	1.092	1.056
t0flfn	0.995	0.939	0.999	0.920	1.009	0.902	0.976	0.867
t0fncn	1.867	1.867	1.867	1.867	1.867	1.867	2.144	2.144
e0olon	1.058	1.015	1.094	1.013	1.134	1.017	1.069	0.946
f0olon	1.058	1.015	1.094	1.013	1.134	1.017	1.069	0.946
Xolon	1.073	1.101	1.117	1.168	1.167	1.240	1.099	1.147
Uolon	0.996	0.996	0.995	0.995	0.997	0.995	0.991	0.990
Volon	0.996	0.996	0.995	0.995	0.997	0.995	0.991	0.990
Wolon	0.996	0.996	0.995	0.995	0.997	0.995	0.991	0.990
e0flfn	0.995	0.939	0.999	0.920	1.009	0.902	0.976	0.867
f0flfn	0.995	0.939	0.999	0.920	1.009	0.902	0.976	0.867
Xflfn	1.054	1.089	1.088	1.147	1.128	1.206	1.065	1.139
Uflfn	0.999	0.999	1.0	0.999	1.002	1.000	0.999	0.993
Vflfn	0.999	0.999	1.0	0.999	1.002	1.000	0.999	0.993
Wflfn	0.999	0.999	1.000	0.999	1.002	1.000	0.999	0.993
e0clcn	1.092	1.092	1.157	1.132	1.243	1.171	1.092	1.056
f0clcn	1.092	1.092	1.157	1.132	1.243	1.171	1.092	1.056
Xclcn	1.130	1.130	1.221	1.223	1.341	1.343	1.143	1.146
Uclcn	1.009	1.009	1.015	1.015	1.023	1.022	1.007	1.007
Vclcn	1.009	1.009	1.015	1.015	1.023	1.022	1.007	1.007
Wclcn	1.009	1.009	1.015	1.015	1.023	1.022	1.007	1.007
Woncn	0.972	0.972	0.972	0.972	0.972	0.972	0.974	0.976
Wolcl	0.960	0.960	0.954	0.953	0.948	0.946	0.958	0.958
Wfncn	0.958	0.958	0.958	0.958	0.958	0.958	0.954	0.954
Wflcl	0.948	0.949	0.944	0.943	0.939	0.937	0.942	0.941

Table 3.3b

pars/vars	case 5	case 5	case 6	case 6	case 7	case 7	case 8	case 8
	REV	IRREV	REV	IRREV	REV	IRREV	REV	IRREV
πa e0*	0.5	80.0	0.5	80.0	0.5	80.0	0.5	80.0
πb e1*	0.5	80.0	0.5	80.0	0.5	80.0	0.5	80.0
γ f0*	0.0	80.0	0.0	80.0	1.0	80.0	-1.0	80.0
$\bar{a}$ f1*	0.75	80.0	0.75	80.0	0.75	80.0	0.75	80.0
va ρ	0.3	0.95	0.5	0.95	0.3	0.95	0.3	0.95
$\bar{b}$ δ	0.75	0.9	0.75	0.9	0.75	0.9	0.75	0.9
vb λ	0.3	0.5	0.5	0.5	0.3	0.5	0.3	0.5
t0olcn	1.634	1.356	1.804	1.323	1.808	1.808	1.469	1.040
t0olcl	1.409	1.268	1.302	1.146	1.362	1.639	1.469	1.040
t0clcn	1.159	1.069	1.385	1.155	1.327	1.103	1.000	1.000
t0olon	1.112	0.923	1.228	0.901	1.231	1.231	1.000	0.708
t0oncn	1.469	1.469	1.469	1.469	1.469	1.469	1.469	1.469
t0flcn	2.082	1.766	2.119	1.580	2.430	1.863	1.766	1.308
t0flcl	1.796	1.652	1.530	1.369	1.831	1.689	1.766	1.308
t0clcn	1.159	1.069	1.385	1.155	1.327	1.103	1.000	1.000
t0flfn	0.971	0.824	0.989	0.737	1.133	0.869	0.824	0.610
t0fncn	2.144	2.144	2.144	2.144	2.144	2.144	2.144	2.144
e0olon	1.112	0.923	1.228	0.901	1.231	1.231	1.000	0.708
f0olon	1.112	0.923	1.228	0.901	1.231	1.231	1.000	0.708
Xolon	1.149	1.232	1.302	1.431	1.305	1.305	1.000	1.142
Uolon	0.988	0.986	0.988	0.978	1.030	1.030	0.949	0.952
Volon	0.988	0.986	0.988	0.978	1.030	1.030	0.949	0.952
Wolon	0.988	0.986	0.988	0.978	1.030	1.030	0.949	0.952
e0flfn	0.971	0.824	0.989	0.737	1.133	0.869	0.824	0.610
f0flfn	0.971	0.824	0.989	0.737	1.133	0.869	0.824	0.610
Xflfn	1.109	1.210	1.231	1.363	1.289	1.193	0.945	1.111
Uflfn	0.993	0.990	0.997	0.987	1.035	1.050	0.956	0.958
Vflfn	0.993	0.990	0.997	0.987	1.035	1.050	0.956	0.958
Wflfn	0.993	0.990	0.997	0.987	1.035	1.050	0.956	0.958
e0clcn	1.159	1.069	1.385	1.155	1.327	1.103	1.000	1.000
f0clcn	1.159	1.069	1.385	1.155	1.327	1.103	1.000	1.000
Xclcn	1.248	1.252	1.601	1.592	1.510	1.511	1.000	1.000
Uclcn	1.012	1.012	1.029	1.028	1.025	1.024	1.000	1.000
Vclcn	1.012	1.012	1.029	1.028	1.025	1.024	1.000	1.000
Wclcn	1.012	1.012	1.029	1.028	1.025	1.024	1.000	1.000
Woncn	0.974	0.974	0.974	0.974	0.974	0.974	0.974	0.974
Wolcl	0.951	0.949	0.935	0.927	0.979	0.980	0.924	0.927
Wfncn	0.954	0.954	0.954	0.954	0.954	0.954	0.954	0.954
Wflcl	0.936	0.934	0.924	0.914	0.963	0.978	0.912	0.914

Table 3.3c

pars/vars	case 9	case 9	case 10	case 10	case 11	case 11	case 12	case 12
	REV	IRREV	REV	IRREV	REV	IRREV	REV	IRREV
πa e0*	0.5	80.0	0.5	80.0	0.5	80.0	0.3	80.0
πb e1*	0.5	65.0	0.5	100.0	0.5	80.0	0.7	80.0
γ f0*	0.0	80.0	0.0	80.0	0.0	80.0	0.0	80.0
$\bar{a}$ f1*	0.75	65.0	0.75	100.0	0.75	80.0	0.75	80.0
va ρ	0.3	0.95	0.3	0.95	0.1	0.95	0.1	0.95
$\bar{b}$ δ	0.75	0.9	0.75	0.9	0.75	0.9	0.75	0.9
vb λ	0.3	0.5	0.3	0.5	0.5	0.5	0.5	0.5
t0olcn	1.429	1.205	2.292	1.880	1.626	1.362	1.574	1.410
t0olcl	1.290	1.297	1.731	1.419	1.418	1.271	1.433	1.320
t0clcn	1.108	0.929	1.324	1.324	1.146	1.072	1.098	1.068
t0olon	1.085	0.894	1.172	0.961	1.197	0.928	1.072	0.961
t0oncn	1.317	1.348	1.956	1.956	1.496	1.469	1.467	1.467
t0flcn	1.732	1.320	3.207	2.706	2.041	1.702	2.140	1.852
t0flcl	1.564	1.422	2.421	2.043	1.781	1.588	1.916	1.734
t0clcn	1.108	0.929	1.324	1.324	1.146	1.072	1.098	1.068
t0flfn	0.976	0.727	0.962	0.812	0.952	0.794	0.961	0.846
t0fncn	1.774	1.815	3.332	3.332	2.144	2.144	2.190	2.190
e0olon	1.085	0.894	1.172	0.961	0.857	0.665	0.928	0.808
f0olon	1.085	0.894	1.172	0.961	1.357	1.191	1.262	1.161
Xolon	1.149	1.254	1.149	1.209	1.141	1.221	1.097	1.150
Uolon	0.988	0.982	0.988	0.988	0.924	0.911	0.954	0.940
Volon	0.988	0.982	0.988	0.988	1.039	1.049	1.022	1.034
Wolon	0.988	0.982	0.988	0.988	0.981	0.980	0.987	0.986
e0flfn	0.976	0.727	0.962	0.812	0.770	0.361	0.860	0.549
f0flfn	0.976	0.727	0.962	0.812	1.135	1.228	1.084	1.208
Xflfn	1.109	1.182	1.109	1.188	1.097	1.188	1.066	1.131
Uflfn	0.993	0.994	0.993	0.993	0.924	0.876	0.953	0.912
Vflfn	0.993	0.994	0.993	0.993	1.049	1.085	1.031	1.070
Wflfn	0.993	0.994	0.993	0.993	0.987	0.980	0.991	0.988
e0clcn	1.108	0.929	1.324	1.324	1.146	1.072	1.098	1.068
f0clcn	1.108	0.929	1.324	1.324	1.146	1.072	1.098	1.068
Xclcn	1.248	1.250	1.248	1.248	1.228	1.232	1.156	1.159
Uclcn	1.012	1.009	1.012	1.012	0.994	0.995	0.996	0.996
Vclcn	1.012	1.009	1.012	1.012	1.028	1.027	1.019	1.018
Wclcn	1.012	1.009	1.012	1.012	1.011	1.011	1.007	1.007
Woncn	0.975	0.975	0.974	0.974	0.974	0.974	0.974	0.974
Wolcl	0.952	0.949	0.950	0.951	0.945	0.944	0.954	0.953
Wfncn	0.956	0.956	0.954	0.954	0.954	0.954	0.953	0.953
Wflcl	0.936	0.942	0.935	0.936	0.931	0.925	0.937	0.935

Table 3.3d

pars/vars	case 13	case 13	case 14	case 14	case 15	case 15	case 16	case 16
	REV	IRREV	REV	IRREV	REV	IRREV	REV	IRREV
πa e0*	0.5	80.0	0.5	80.0	0.5	80.0	0.5	80.0
πb e1*	0.5	80.0	0.5	80.0	0.5	80.0	0.5	80.0
γ f0*	0.0	100.0	0.0	100.0	0.0	100.0	-1.0	100.0
$\bar{a}$ f1*	0.75	100.0	0.75	100.0	0.75	100.0	0.75	100.0
va ρ	0.3	0.95	0.1	0.95	0.5	0.95	0.5	0.95
$\bar{b}$ δ	0.75	0.9	0.75	0.9	0.75	0.9	0.75	0.9
vb λ	0.3	0.5	0.5	0.5	0.1	0.5	0.1	0.5
t0olcn	1.634	1.376	1.626	1.426	1.626	1.350	1.503	1.131
t0olcl	1.409	1.273	1.418	1.325	1.418	1.254	1.461	1.131
t0clcn	1.159	1.081	1.146	1.077	1.146	1.077	1.029	1.001
t0olon	1.112	0.914	1.107	0.949	1.107	0.898	1.023	0.752
t0oncn	1.469	1.504	1.469	1.504	1.469	1.504	1.469	1.504
t0flcn	2.082	1.737	2.041	1.771	2.041	1,715	1.811	1.423
t0flcl	1.796	1.607	1.781	1.645	1.781	1.593	1.760	1.422
t0clcn	1.159	1.081	1.146	1.077	1.146	1.077	1.029	1.001
t0flfn	0.971	0.791	0.952	0.807	0.952	0.781	0.845	0.648
t0fncn	2.144	2.195	2.144	2.195	2.144	2.195	2.144	2.195
e0olon	1.170	0.771	0.783	0.422	1.540	1.137	1.377	0.827
f0olon	1.084	0.986	1.266	1.208	0.893	0.780	0.849	0.716
Xolon	1.149	1.233	1.141	1.228	1.141	1.214	1.031	1.145
Uolon	0.983	0.973	0.895	0.871	1.053	1.065	1.007	1.015
Volon	0.991	0.991	1.030	1.039	0.942	0.933	0.924	0.923
Wolon	0.987	0.983	0.967	0.961	0.994	0.994	0.963	0.965
e0flfn	0.963	0.851	0.700	0.142	1.176	1.239	1.009	0.948
f0flfn	0.977	0.754	1.109	1.221	0.813	0.497	0.742	0.462
Xflfn	1.109	1.202	1.097	1.189	1.097	1.186	0.975	1.119
Uflfn	0.990	1.004	0.895	0.826	1.068	1.118	1.022	1.063
Vflfn	0.994	0.983	1.037	1.063	0.942	0.904	0.927	0.901
Wflfn	0.992	0.993	0.971	0.953	1.000	1.003	0.971	0.976
e0clcn	1.317	1.165	1.291	1.156	1.291	1.156	1.057	1.001
f0clcn	1.106	1.053	1.097	1.051	1.097	1.051	1.019	1.000
Xclcn	1.248	1.249	1.228	1.228	1.228	1.228	1.045	1.047
Uclcn	1.017	1.026	0.992	1.006	1.036	1.048	1.019	1.023
Vclcn	1.009	1.003	1.022	1.013	0.996	0.989	0.992	0.990
Wclcn	1.013	1.014	1.008	1.009	1.016	1.017	1.005	1.005
Woncn	0.973	0.972	0.972	0.972	0.973	0.972	0.973	0.972
Wolcl	0.948	0.942	0.934	0.925	0.952	0.950	0.932	0.933
Wfncn	0.952	0.951	0.952	0.951	0.952	0.951	0.952	0.951
Wflcl	0.933	0.931	0.918	0.898	0.938	0.938	0.920	0.923

References

Arrow, K.J. and A.C. Fisher (1974) Environmental Preservation, Uncertainty and Irreversibility. *Quarterly Journal of Economics, 88, pp. 312-319.*

Barrett, S. (1992) International Environmental agreements as Games. In: R. Pethig (Ed.) *Conflicts and Cooperation in Managing Environmental Resources*, Springer-Verlag, Berlin, pp. 11-36.

Beltratti, A., G. Chichilnisky and G. Heal (1993) Preservation, Uncertain Future Preferences and Irreversibility. *Mimeo*, Columbia Business School.

Carraro,C and D. Siniscalco (1991) The International Protection of the Environment: Voluntary Agreements Among Sovereign Countries. *FEEM Working Paper, 1.91.*

Chichilnisky, G. and G. Heal (1993) Global Environmental Risks. *Journal of Economic Perspectives, 7, pp.65 - 86.*

Epstein, L. G. (1980) Decision Making and the Temporal Resolution of Uncertainty. *International Economic Review, 21, pp. 269-283.*

Freixas, X. and J-J. Laffont (1984) On the Irreversibility Effect. In: M. Boyer and R. Kihlstrom (Eds.) *Bayesian Models in Economic Theory*, Ch. 7., Elsevier, Dordrecht.

Heal, G. (1992) International Agreements on Emission Control. *Structural Change and Economic Dynamics, 3, pp. 223 - 240.*

Henry, C. (1974a) Investment Decisions Under Uncertainty: The Irreversibility Effect. *American Economic Review, 64, pp.1006-1012.*

Henry, C. (1974b) Option Values in the Economics of Irreplaceable Assets. *Review of Economic Studies, Symposium on the Economics of Exhaustible Resources, pp. 89 - 104.*

Hoel, M. (1992a) Emission Taxes in a Dynamic International Game of CO_2 Emissions. In: R. Pethig (Ed.) *Conflicts and Cooperation in Managing Environmental Resources*, Springer-Verlag, Berlin, pp. 39-68.

Hoel, M. (1992b) The Role and Design of a Carbon Tax in an International Climate Agreement. In: *Climate Change: Designing a Practical Tax System*, OECD, Paris.

Hoel, M. (1992c) *How Should International Greenhouse Gas Agreements Be Designed?* In: P. Dasgupta, K-G Maler and A. Vercelli (Eds.) *The Economics of Transnational Commons*, OUP, Oxford (forthcoming).

Kolstad, C. (1992) Regulating a Stock Externality under Uncertainty with Learning. *Mimeo,* University of Illinois.

Kolstad, C. (1993a) Looking vs Leaping: The Timing of CO_2 Control in the Face of Uncertainty and Learning. *Mimeo*, University of Illinois.

Kolstad, C. (1993b) The Implications of Learning About Uncertainty for US/European CO_2 Control. In: J. Braden, H. Folmer and T. Ulen (Eds.) *Environmental Policy with Economic and Political Integration: the EC and the US*, Edward Elgar (forthcoming).

Kolstad, C. (1993c) Mitigating Climate Change Impacts: The Conflicting Effects of Irreversibility in CO_2 Accumulation and Emission Control Investment. *Paper presented to IIASA Workshop on Integrative Assessment of Mitigation,Impacts and Adaptation to Climate Change*, IIASA, Laxenburg.

Maler, K-G. (1990) Incentives in International Environmental Problems. In: H. Siebert (Ed.) *Environmental Protection : The International Dimension*, J.C.B. Mohr, Tubingen.

Manne, A, and R. Richels (1992) *Buying Greenhouse Gas Insurance: The Economic Costs of CO_2 Emission Limits.* MIT Press, Cambridge, Mass.

Peck, S. and T. Tiesberg (1992) Global Warming Uncertainties and the Value of Information: An Analysis Using CETA. *Mimeo,* EPRI.

Ulph, A. and D. Ulph (1994a) Global Warming: Why Irreversibility May Not Require Lower Current Emissions of Greenhouse Gases. *University of Southampton Discussion Paper No. 9402.*

Ulph, A. and D. Ulph (1994b) The Irreversibility Effect Revisited. *Mimeo,* University of Southampton.

Ploeg, van der R. and A. de Zeeuw (1992) International Aspects of Pollution Control. *Environmental and Resource Economics , 2, pp. 117 - 140.*

4 Voluntary Supply of Greenhouse Gas Abatement and Emission Trading Equilibria

Johan Eykmans and Stef Proost*
Faculteit Economische en Toegepaste Economische Wetenschappen
Centrum voor Economische Studien
Katholieke Universiteit Leuven
Naamsestraat 69
B-3000 Leuven, Belgium

Abstract

This paper examines equilibria for a global pollution problem where the level of emission reduction is determined by voluntary and noncooperative supply of abatement by different countries. Starting from the Nash Cournot equilibrium, we give each country the possibility to buy or sell additional emission reduction at a fixed price in an international abatement market. We show in this paper that in a competitive market for emission abatement, the country with the highest willigness to pay for environmental quality determines the total level of emission reduction. It is also shown that this player is not necessarily better off under the competitive trading scenario compared to the Nash Cournot equilibrium without trade. In an empirical illustration the Nash Cournot sollution without trade is compared to a competitive and monopsonistic emission abatement trading equilibrium. In the monopsonistic trading case we distinguish between no price discrimination and perfect price discrimination. We conclude that some form of price discrimination by the monopsonistic is a necessary, though not a sufficient, requirement to achieve a Pareto improvement compared to the Nash Cournot equilibrium without trade.

---

* Johan Eykmans is a research assistant of the Belgian National Fund for Scientific Research NFWO and Stef Proost is research fellow of the same fund. Support from the research program "Energy and the Environment: Markets and Policies" of the Fondazione ENI Enrico Mattei is gratefully acknowledged. We would like to thank Micheal Hoel, Erik Schokkaert and an anonymous referee for comments on an earlier version of this paper.

NATO ASI Series, Partnership Sub-Series, 2. Environment – Vol. 14
Economics of Atmospheric Pollution
Edited by Ekko C. van Ierland and Kazimierz Górka

4.1 Introduction

The greenhouse problem can be considered as an externality problem on world scale. Emissions of greenhouse gases (of which CO_2 is the principal anthropogenetic gas) from all countries mix uniformly in the Earth's atmosphere and will have climate change repercussions all over the world. Theoretically speaking, global international agreements are the correct response to this type of environmental externalities. The merits of a great variety of alternative agreement structures have been studied during the last years, for a survey see a.o. Barrett (1992a). There have been proposals involving uniform emission reduction efforts like in Hoel (1992a) or a system of tradable international carbon emission entitlements, see Barrett (1992b). International carbon taxes with redistribution of the proceeds and minimum national carbon taxes are studied in, e.g., Hoel (1992b) or Eyckmans, Proost and Schokkaert (1994). Although such agreements bring about large cooperative gains, the experience of the Rio and Berlin climate conferences demonstrates that reaching a binding and effective agreement is not evident.

In contrast to the usual assumption that countries can negotiate and enforce binding cooperative agreements we focus in this paper on a noncooperative solution to the greenhouse negotiation problem. In particular we start from the observation that in a Nash Cournot laissez faire equilibrium countries abate their emissions up to the point where their marginal reduction costs equal corresponding marginal abatement benefits. Because the latter benefits are likely to differ between countries, marginal abatement costs are not equalized and cost inefficiencies in the production of emission abatement will emerge (productive inefficiency). Moreover, a second type of inefficiency occurs because the total level of abatement is too low since each country neglects the benefits to other countries of its own emission reduction efforts (allocative inefficiency). However, we will focus in this paper on the productive efficiency aspect and explore a mechanism that achieves Pareto improvements over the Nash Cournot laissez faire outcome.

The first idea we explore, is the creation of an artificial international market for the environmental good greenhouse gas emission abatement. For a given price per unit of abatement, every individual country determines its optimal supply of and demand for emission abatement. Excess supply and demand can be traded on the international market for abatement and the price is assumed to adjust until this market clears. In this way we can alleviate the productive inefficiency mentioned above. The trading mechanism could take different institutional forms. We can think of a system where countries with high marginal benefits from abatement make bilateral deals with countries characterized by low costs of abatement. Alternatively we can think of a central agency organizing a market for emission reduction. The auctioneer sets a price and in function of the resulting demand and supply of abatement, the price is appropriately adjusted. Finally, we might think of an environmental fund, created by one country, that invites all other countries to make an offer to produce emission reduction.

In section 4.3 the notion of a noncooperative Competitive Emission Trading Equilibrium is introduced. This is an equilibrium with noncooperative supply of abatement and with competitive trade in emission reduction. In a first step we assume that trade in emission reduction is competitive, i.e., market participants are assumed to be negligibly small with respect to the total trading volume implying that they take the price level as given. For quasi linear preferences over income and joint emission abatement we show in section 4.4 that only one unique country, characterized by the highest willingness to pay for abatement, will determine the total level of emission reduction. This particular player chooses to buy abatement from low cost countries instead of producing all emission reduction by himself. Total abatement and total pay off are higher under the competitive trading scenario than in the noncooperative Nash Cournot Equilibrium Without Trade. In terms of utility all sellers of abatement are better off if trading is possible. Only the unique purchaser of emission reduction enjoys less pay off under competitive emission reduction trading than in the laissez faire equilibrium. This is a disturbing conclusion because it makes the trading mechanism hard to swallow for the country with highest willingness to pay.

Following Hahn (1984), we relax in section 4.5 the competitive trade assumption and investigate how a single emission abatement buyer can use its market power with and without price discrimination. In a monopsonistic market setting, the purchaser of abatement can lower the international price of emission reduction by restraining his own demand or by increasing his domestic production of abatement effort. If he cannot discriminate between the different suppliers (i.e., if he cannot charge individualized prices) he is very likely to be worse off with an emission trading scheme than without it. Only if the purchaser can reap all producers' surpluses he might gain compared to the noncooperative Nash Cournot Equilibrium Without Trade. Hence, in order to make all parties accept a voluntary agreement to install an international emission abatement trading mechanism, it will be necessary to reimburse some of the trade gains from the sellers to the purchaser of abatement. Alternatively, some kind of individualized price system for emission abatement is required if the trading mechanism is to be installed through a voluntary international agreement. Unfortunately, it turns out to be impossible to prove analytically the claim that the monopsonist would gain compared to the noncooperative Nash Cournot Equilibrium Without Trade if he can perfectly discriminate against all suppliers of abatement. The sign of this change in welfare depends crucially upon the relative slopes of marginal cost and benefits functions and has to be investigated by means of simulations.

In section 4.6 we illustrate the emission trading solutions using a simple model for the world greenhouse problem. We distinguish between 12 regions and calibrate logarithmic cost and quadratic benefit functions to estimates available in the literature. The cost functions are the same as in Eyckmans, Proost and Schokkaert (1993), the marginal benefits however are chosen to be linear functions of joint abatement effort. Simulations confirm the analytical results of section 4.4 and 4.5. The simulations lead to the conclusion that, for the parameter values chosen, the monopsonistic emission trading equilibrium with perfect price discrimination establishes a Pareto improvement over the noncooperative laissez faire outcome. Section 4.7 concludes.

4.2 Preferences, noncooperative and cooperative equilibria

Consider the case of greenhouse gas emission abatement. $N=\{1,2,\ldots,3\}$ is the set of countries in the world. At the one hand, reducing its emissions of carbon dioxide inflicts a cost upon country i in terms of forgone output since emissions are to be considered as a necessary input in production. On the other hand, lower emissions of carbon dioxide give rise to benefits because expected damage of climate change is reduced. We use a quasi linear utility function to represent individual countries' preferences over income times total emission abatement[2]:

$$u_i(r_i, r_N) = B_i(r_N) - C_i(r_i) \qquad i \in N \qquad r_N = \sum_{j \in N} r_j \tag{4.1}$$

[2] Utility is assumed to be linear in income implying that marginal valuation of income is identical and equal to unity for all players in the game. Hence, income effects will not play. For the representation in (4.1) we also assumed that all players in the game are approximately of the same size in terms of population and initial endowments. However, the analysis can easily be extended to accommodate differences in these variables.

Individual abatement r_i is defined over a closed interval $[0, E_i^0]$ with E_i^0 some strictly positive finite upper bound on individual emission abatement[3]. An allocation of emission abatement in the greenhouse game is a *n*-tuple $(r_1, r_2, \ldots, r_n)$ and is denoted by *r*. Joint abatement r_N is simply the sum of individual emission reduction efforts and is to be considered as a nonexcludable public good. First, no country can be excluded from the benefits of total emission reduction, and secondly, consumption of this abatement involves nonrivalry. Standard assumptions on the cost and benefit of abatement functions over their corresponding domains are taken for granted:

$$C_i'(0) = 0, \qquad C_i'(r_i) > 0 \quad for \quad r_i > 0, \tag{4.2}$$

$$\lim_{r_i \to E_i^0} C_i'(r_i) = +\infty, \qquad C_i''(r_i) \geq 0$$

Abatement costs are a strictly increasing and convex function of individual emission reduction effort. The first unit of abatement is assumed to be free, abating the last unit of emissions, however, is infinitely costly. Benefits of abatement are strictly increasing and concave in total emission reduction.

$$B_i'(r_N) > 0, \qquad B_i'(0) = \theta_i \in \mathrm{IR}_{++}, \qquad B_i''(r_N) \leq 0 \tag{4.3}$$

The latter condition just requires that marginal benefits of the first unit of abatement is strictly positive and finite for all *i* in N. If no agreement on cooperation in the reduction of emissions is reached, every player will only take into account the benefits of abatement which accrue to himself. All other countries' emission abatement is considered as exogenously given. Maximizing individual utility under the Nash behavioural assumption yields the familiar first order condition for a noncooperative Nash Cournot Equilibrium Without Trade[4] (NCEWT) denoted r^{nc}:

$$C_i'(r_i^{nc}) = B_i'(r_N^{nc}) \qquad (\forall i \in N) \tag{4.4}$$

[3] It is commonly assumed in this kind of emission abatement models that abatement is always positive for all countries involved, see Hoel (1992a,b). Hence emissions cannot be increased compared to the reference year. This assumption is made only for convenience and is not crucial to the results in this paper. By choosing another reference emission level $\hat{E}_i^0$ negative abatement levels are possible for some countries. This can be interpreted as saying that, for these particular countries, there is a opportunity cost element to reducing their emissions.

[4] The assumptions on cost and benefit functions guarantee the existence of at least one Nash Cournot Equilibrium Without Trade.

Every country reduces its emissions up to the point where its individual marginal abatement cost equals its individual marginal benefit from total emission reduction. The latter marginal benefit of abatement is a measure for the willingness to pay of country *i* for the provision of the public good r_N or for environmental quality in general. Inspection of the conditions in (4.4) tells us that marginal abatement costs are probably not equalized if countries value the public good differently in the Nash Cournot Equilibrium Without Trade. This implies that the distribution of abatement is cost inefficient, the same overall emission reduction r_N^{nc} could have been achieved at a lower cost. Moreover, it is well-known that the public good is underprovided in the Nash Cournot Equilibrium Without Trade compared to the first best efficient distribution of abatement efforts r_N^* defined by Samuelson's condition:

$$C_i'(r_i^*) = \sum_{j \in N} B_j'(r_N^*) \qquad (\forall i \in N) \tag{4.5}$$

Conditions (4.5) entail on the one hand cost efficiency because marginal costs are equalized (productive efficiency), and on the other hand an optimal provision of the public good since the positive reciprocal externalities from abatement are appropriately taken into account (allocative efficiency). The problem with this first best allocation of abatement efforts is that some countries might actually end up worse off in utility terms compared to the Nash Cournot Equilibrium Without Trade if lump sum transfers are not available. Therefore, it will be very difficult to implement this first best outcome since nonaltruistic and sovereign nations are not likely to accept an agreement which makes them worse off than without international coordination of greenhouse policies. Given that an agreement to cut back greenhouse gas emissions has to result from voluntary cooperation, it make sense to require that a proposed scheme on abatement efforts makes every participant at least as well off as in the laissez faire noncooperative equilibrium. Hence, the proposal should constitute a Pareto improvement over the Nash Cournot Equilibrium Without Trade. In the sequel of the paper we will say that an allocation *r* of abatement efforts satisfies the *Voluntary Participation Constraint* if and only if

$$u_i(r_i, r_N) \geq u_i(r_i^{nc}, r_N^{nc}) \qquad (\forall i \in N) \tag{4.6}$$

Obviously, the latter is only a necessary, and not a sufficient condition for voluntary agreements on emission abatement to emerge.

4.3 Noncooperative solution with competitive emission trading

In this paper we will not seek to design a procedure yielding a full first best allocation of emission reduction satisfying both allocative and productive efficiency. We will focus on the productive efficiency aspect, i.e., exploiting the differences in marginal abatement costs in order to achieve a Pareto improvement compared to the laissez faire noncooperative solution. For that purpose, we will create an artificial international market for greenhouse gas emission abatement where individual countries can buy or sell emission reduction at a uniform price. Buying additional emission reduction instead of cutting back its domestic emissions might be beneficial to countries characterized by high marginal abatement costs. On the other hand, countries which can reduce their emissions at a relatively low cost might gain from reducing their greenhouse gas emissions by an extra amount and selling this excess supply to high cost nations. Given that such an international trading scheme would exist, the relevant question is whether all market participants end up at least as well off as in the Nash Cournot Equilibrium Without Trade. This question will be investigated under different assumption on the organisation of the emission abatement market.

Assume that countries can buy or sell emission abatement at a fixed price p per unit in an international market for emission reduction. In a first step we assume that the different countries take this international market price as given, this is the *Competitive Emission Trading* case. Each country i chooses to supply effective emission abatement efforts (capital R_i) and to demand emission abatement (lower-class r_i), taking as given the demand for emission abatement by the other countries $r_{-i} = \sum_{j \neq i} r_j$. The basic idea behind this emission trading scheme is to separate the production of the public good on the one hand, and the problems related to revealing the appropriate demand for the public good on the other hand. Notice that this trading scheme is different from the usual emission entitlements markets in environmental economics. Traditionally it is assumed that some central agency decides on the amount of emission abatement needed and issues emission entitlements to achieve that target. The permits are allocated in some way to the players who can freely trade the entitlements among each other. In our trading mechanism total emission abatement is determined endogenously by equating the demand and supply for emission reduction. No agency is set up to issue and allocate emission entitlements. Individual utility can now be written as a function of individual supply and demand of emission and of all other countries' demand[5]:

$$u_i(R_i, r_i, r_{-i}) = B_i(r_i + r_{-i}) - C_i(R_i) + p[R_i - r_i] \qquad (\forall i \in N) \quad (4.7)$$

The term $p[R_i - r_i]$ stands for the profit of selling the surplus of permits if $R_i > r_i$ or the cost of acquiring sufficient permits if $R_i < r_i$. The supply of emission abatement and the choice of the preferred level of provision of the public good are separated in this formulation.

Supply of emission abatement

Country *i* chooses to supply a nonnegative amount of emission reduction $R_i \geq 0$ such that (4.7) is maximized. The corresponding first order condition reads:

$$C'_i(R_i) = p \qquad (\forall i \in N) \quad (4.8)$$

These conditions must hold with equality[6] for all countries and they implicitly define emission abatement supply functions $R_i = R_i(p)$ for all regions and $R_i(p)$ stands for the inverse of the marginal cost function of *i*. Because the cost function is assumed strictly convex in $R_i(p)$, marginal costs are monotonically increasing so that existence of its inverse is guarantied. Because of the convexity of $C_i(R_i)$ it is straightforward to show that $R_i(p)$ increases in *p*. The higher the international price of abatement, the more player *i* wants to produce of it.

Demand for emission abatement

Maximizing (4.7) w.r.t. $r_i \geq 0$ taking as given demand by all other regions r_{-i} yields following Kuhn-Tucker conditions for individual abatement demand:

$$-p + B_i'(r_i + r_{-i}) \leq 0 \quad , \quad r_i \geq 0 \quad , \quad (4.9)$$

$$[-p + B_i'(r_i + r_{-i})]r_i = 0 \qquad (\forall i \in N)$$

[5] Parenthesis are used to denote arguments of functions *f(x)* whereas square brackets are used in algebraic expressions *[1+f(x)]/x*.

[6] The first order conditions in (4.8) hold with equality because we assumed in the beginning that $C_i(R_i)$ is convex and $C'_i(0) = 0 \;\; \forall i \in N$ implying that for any strictly positive price level *p*, it is impossible that $C'_i(R_i) > p \quad \forall R_i \in [0, E_i^0]$.

For positive levels of emission reduction demand country i equates its marginal willingness to pay for abatement, or alternatively its marginal benefits of abatement, to the international price of emission reduction. Condition (4.9) implicitly defines the reaction curve for country i in function of the price p and other countries' emission reduction r_{-i}. For any price level and effort level of the other countries, the reaction function gives the best demand reply of i. Defining $\beta_i(p) \equiv [B_i'(r_N)]^{-1}$ as inverse of i's marginal benefit function, the best reply function for i reads:

$$r_i = \beta_i(p) - r_{-i} \geq 0 \qquad (\forall i \in N) \tag{4.10}$$

Under the standard assumptions concerning $B_i(r_N)$ it can be shown that these reaction curves are downward sloping in both p and r_{-i}. In the next section we will show that the n first order conditions defined by the reaction curves in (4.10) can be solved explicitly for exactly one unique equilibrium demand for abatement which varies continuously with the price p. We now have all elements to define a noncooperative allocation of the abatement efforts allowing for competitive trade. We will call this equilibrium a noncooperative *Competitive Emission Trading Equilibrium* (CETE).

Equilibrium on the competitive market for emission abatement

Definition 1: *A noncooperative Competitive Emission Trading Equilibrium (CETE) consists of a vector of abatement supply* $\tilde{R} \in \times_{j \in N}[0, E_j^0]$, *a vector of emission reduction demand* $\tilde{r} \in \times_{j \in N}[0, E_N^0]$, *and a unique price* $\tilde{p} \in \Re_{++}$ *such that:*

(i) $(\forall i \in N): \; C'_i(\tilde{R}_i) = \tilde{p}$

(ii) $(\forall i \in N)(\forall r_i \in [0, E_N^0]): \; u_i(\tilde{R}_i, \tilde{r}_i, \tilde{r}_{-i}) \geq u_i(\tilde{R}_i, r_i, \tilde{r}_{-i})$

(iii) $\sum_{j \in N} \tilde{R}_j = \sum_{j \in N} \tilde{r}_j$

Conditions *(i)* implies that total emission abatement is produced in a cost effective way (suppliers take the price as given), and *(ii)* requires that the vector of emission demand $\tilde{r}$ is a noncooperative equilibrium. The last condition clears the emission trading market. We do not present here a formal proof of existence for the noncooperative Competitive Emission Trading Equilibrium, but given quasi linearity of preferences, the convexity and concavity assumptions about cost and benefit functions and adding the appropriate smoothness conditions, the conditions for applying Brouwer's fixed point theorem are fulfilled and existence of an equilibrium follows straightforward. Figure 4.1 illustrates the Competitive Emission Trading Equilibrium (CETE) for a two country example.

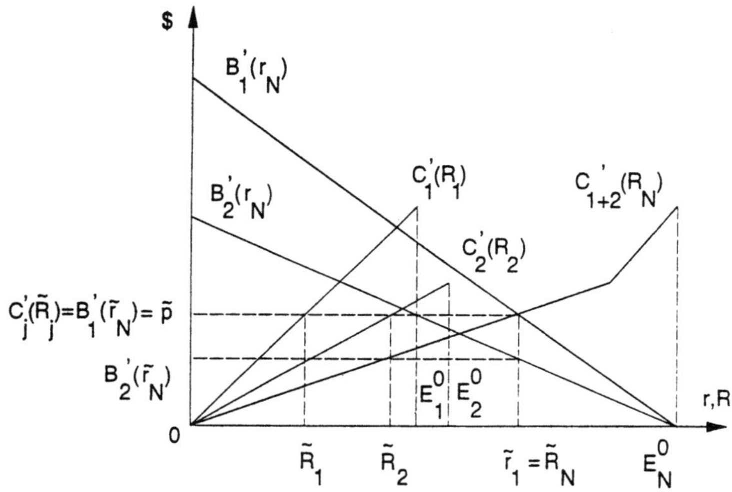

Figure 4.1: Competitive Emission Trading Equilibrium (CETE)

Consider two countries involved in some pollution control problem. Marginal abatement cost and benefit functions are depicted in Figure 4.1. We assume that country *1* is characterized by a lower reference level of emissions $E_1^0 < E_2^0$ and by higher marginal abatement costs than country *2*. At the same time, the marginal willingness to pay of country *1* for emission reduction always lays above country *2*'s willingness to pay. We might think of country *1* as the European Community and country *2* as the former socialist Eastern European countries. Typically, marginal abatement costs are lower in East European countries than in the European Community, however, the latter is willing to pay more for an improvement in environmental quality. The curve denoted by $C_{1+2}'(R_N)$ is the horizontal sum of the individual marginal cost functions. Assuming cost efficient abatement production, this curve gives the aggregate marginal cost of abating total emissions by an amount R_N. The allocation $(\tilde{R}_1, \tilde{R}_2, \tilde{r}_1, 0, \tilde{p})$ is the Competitive Emission Trading Equilibrium CETE as defined in Definition 1. For the price level $\tilde{p}$ marginal abatement costs are equalized implying cost efficient production of emission reduction. Referring to first-order conditions (4.9), country *1* is the only buyer of abatement and equates its marginal benefit to the price level. Country *2*'s marginal benefit falls short of the price level $\tilde{p}$ and hence, it does not buy additional abatement. The latter observation is fomalized in Proposition 1.

4.4 Some properties of the noncooperative Competitive Emission Trading Equilibrium

For a quasi linear specification of preferences, a remarkable property of this competitive emission trading mechanism emerges. Assuming that all players can be ranked strictly according to their willingness to pay for the public good, it is easy to see that only one players will buy emission abatement, namely the player with highest marginal benefit of abatement. The requirement that players can be ranked in function of marginal benefits is equivalent to saying that marginal benefit functions do not intersect over their respective domains.

Proposition 1: *Assume all players can be ranked in strictly decreasing order of their marginal benefits of abatement. There is one and only one player who demands and buys emission reduction in the noncooperative Competitive Emission Trading Equilibrium CETE, namely the player with the highest marginal benefit.*

Proof: Assume players can be ranked and are indexed in strictly decreasing order of marginal benefits:

$$B_1'(r_N) > B_2'(r_N) > \dots > B_n'(r_N) \qquad (\forall r_N \in \Re_+)$$

(i) Because $B_1'(0) > 0$ there will always exist some strictly positive price level $p \leq B_i'(0)$ such that player *1* buys a strictly positive amount of emission reduction.

(ii) Consider the case where both player *1* and *2* are demanding a strictly positive amount of emission abatement. From first order condition (4.10) this implies that $B_1'(\tilde{r}_N) = p = B_2'(\tilde{r}_N)$ which contradicts the ranking assumption.

One might wonder whether it is a strong assumption to make that marginal benefits curves do not intersect. First, it should be noted that marginal benefits of emission abatement reflect the marginal environmental damage caused by emissions. Clearly, countries differ strongly in geographical and climatological characteristics and hence they are likely to differ in their vulnerability to environmental damage caused by particular pollutants. Secondly, although we require marginal benefit functions not to intersect[7], this condition is not crucial to Proposition 1. In the two player example, if marginal benefit curves would intersect, it still holds that only one player will purchase emission abatement in some interval of the emission abatement domain. Ignoring intersection points, we typically obtain a picture where player *1* purchases abatement from *2* for high price levels and vice versa for low prices.

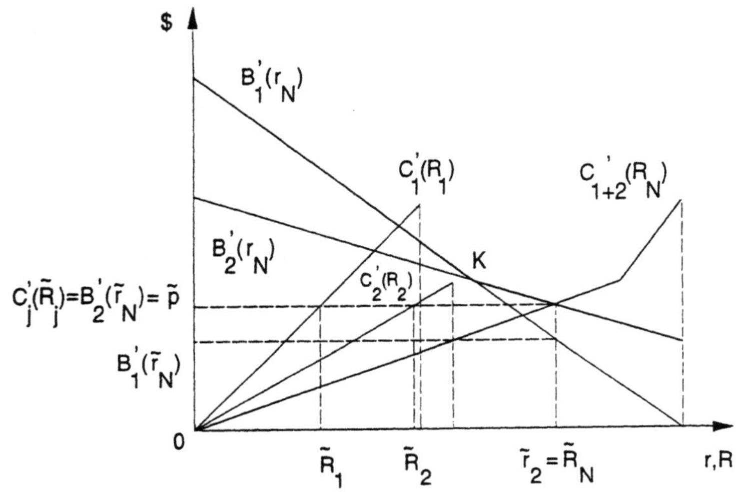

Figure 4.2: CETE and intersecting marginal benefit curves

In Figure 4.2, the marginal benefit curves do intersect at point *K*. Left of *K*, country *1*'s marginal benefits outweigh country *2*'s marginal benefits and the opposite holds true right of *K*. Given this configuration of marginal cost functions, country *2* becomes the only buyer of abatement because at the price level $\tilde{p}$ the aggregate marginal cost function intersects with country *2*'s marginal willingness to pay curve in the region where benefits are higher for *2* than for *1*, i.e., right of *K*. Country *1*'s marginal willingness to pay is strictly less than $\tilde{p}$ in the point $\tilde{R}_N$ and it restrains from purchasing additional emission reduction.Finally, Figure 4.3 illustrates a peculiar case in which both marginal benefit curves and the aggregate marginal cost function intersect in point *K*. If the price level would coincide with the level of marginal

[7] In many applied studies, individual benefits of abatement are assumed to be proportional, i.e., a global benefit function is postulated and a vector of weights is used to distribute benefits over the countries involved: $B_i(r_N) = w_i B(r_N)$ with $w_i > 0 \wedge \sum_{j \in N} w_j = 1$. For this type of specification, the nonintersection assumption always holds.

benefits where both curves intersect, the two agents will demand an equal and strictly positive amount of abatement. At the price $\tilde{p}$ both countries want the same level of abatement $\tilde{R}_N$ which can be produced in a cost efficient way. After trading, there is a net monetary transfer amounting to $\tilde{p}[\tilde{R}_2 - \tilde{R}_1]$ from country *1* towards country *2*.

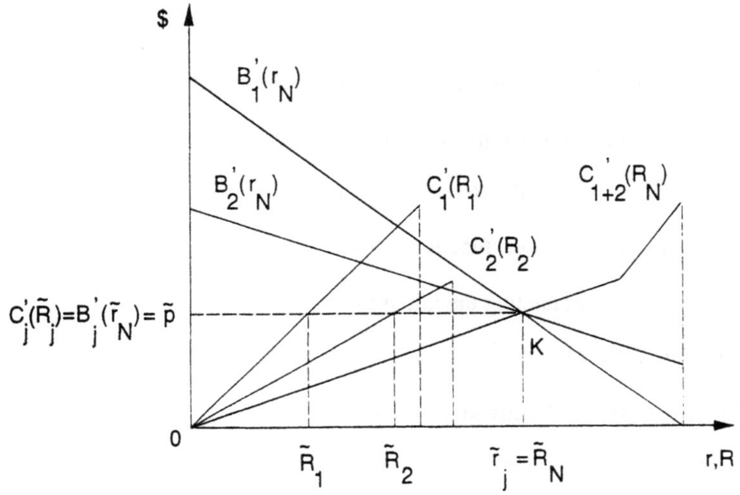

Figure 4.3: A peculiar case of intersection marginal benefit curves

If it is not the nonintersection assumption, which property of the model does drive the result in Lemma 1? We conjecture that the combination of *(i)* quasi linearity of preferences (implying marginal willingness to pay for total abatement to be a function of joint emission reduction only), and *(ii)* the nonexcludable character of the public good total emission abatement, is responsible for the uniqueness result in proposition 1. By definition, differentiation in the level of the public good consumed by the agents is impossible. All have to accept the unique provision level irrespective of their own preferred amount.

What about the level of total emission reduction and individual abatement efforts compared to the Nash Cournot Equilibrium Without Trade (NCEWT)? Proposition 2 shows that allowing for emission trading between the agents in the emission game results in a higher level of total abatement, and the unique purchaser of emission reduction will produce less abatement by himself compared to the NCEWT scenario.

Proposition 2: *(i) Total emission reduction is higher under the noncooperative Competitive Emission Trading Equilibrium CETE than in the corresponding Nash Cournot Equilibrium Without Trade NCEWT.*

(ii) The unique buyer of emission abatement however supplies less emission reduction than in NCEWT.

Proof:(i) Assume the contrary that $\tilde{r}_N < r_N^{nc}$. From the concavity of benefit functions and from the first order conditions for the NCEWT and CETE we can write for all i in N:

$$C_i'(\tilde{R}_i) = p \geq B_i'(\tilde{r}_N) > B_i'(r_N^{nc}) = C_i'(r_i^{nc}) \qquad (\forall i \in N)$$

Hence, from convexity of cost functions $(\forall i \in N)$: $\tilde{R}_i > r_i^{nc}$ which implies that $\tilde{R}_N = \tilde{r}_N > r_N^{nc}$ contradicting the initial assumption.

(ii) Assume the converse $\tilde{R}_1 > r_1^{nc}$. From the convexity of player *1*'s cost function and from the first order conditions we can write:

$$B_1'(\tilde{r}_N) = p = C_1'(\tilde{R}_1) > C_1'(r_1^{nc}) = B_1'(r_N^{nc})$$

Therefore, concavity of the benefit function implies $\tilde{r}_N < r_N^{nc}$ and contradicts the result in *(i)*. Notice that the CETE equilibrium cannot be First Best Pareto efficient because, comparing first order conditions (4.5), (4.8) and (4.9) learns that:

$$C_i'(\tilde{R}_i) = p = B_1'(\tilde{r}_N) \neq \sum_{j \in N} B_j'(r_N^*)$$

Hence, the CETE equilibrium is to be situated in between the noncooperative NCEWT and the First Best Pareto efficient solution in terms of total emission reduction and total surplus. Although total utility is higher if emission trading is allowed for, the solution does not satisfy the Voluntary Participation requirement (4.6) in general because the unique abatement purchaser might lose compared to the laissez faire equilibrium. Therefore, it is unlikely that a competitive emission trading mechanism would result as an institution from voluntary international negotiations. The country with the highest marginal willingness to pay has little incentives to accept an agreement were it is doomed to lose.

Proposition 3: *(i) Total utility under noncooperative CETE is strictly higher than in the corresponding NCEWT equilibrium.*

(ii) All sellers of emission abatement in the CETE are strictly better off than in NCEWT.

(iii) The unique purchaser of abatement might win or lose depending upon the slopes of its marginal benefit and cost functions.

Proof: *(i)* Consider for all *i* in N the following first order approximations of both benefit and cost functions in the Nash Cournot Equilibrium Without Trade:

$$B_i(r_N^{nc}) = B_i(\tilde{r}_N) + B_i'(\tilde{r}_N)[r_N^{nc} - \tilde{r}_N] + \mu_i$$

The residual term is negative because of the concavity of benefit functions:

$$\mu_i = \frac{B_i''(z)}{2}[r_N^{nc} - \tilde{r}_N]^2 \qquad z \in [r_N^{nc}, \tilde{r}_N]$$

Similarly, the cost functions of all players can be approximated with positive residual terms ν_i (because $C_i(R_i)$ are convex):

$$C_i(r_i^{nc}) = C_i(\tilde{R}_i) + C_i'(\tilde{R}_i)[r_i^{nc} - \tilde{R}_i] + \nu_i$$

The difference in total pay off in the NCEWT and CETE equilibrium can now be written as:

$$\sum_{j \in N}[-B_j'(\tilde{r}_N)[r_N^{nc} - \tilde{r}_N] + C_j'(\tilde{R}_j)[r_j^{nc} - \tilde{R}_j] + \tilde{p}[\tilde{R}_j - \tilde{r}_j] - \mu_j + \nu_j]$$

Rearranging and using the first order conditions for CETE one obtains:

$$[\tilde{r}_N - r_N^{nc}]\left[\sum_{j \in N}[B_j'(\tilde{r}_N)] - \tilde{p}\right] + \sum_{j \in N}[\nu_j - \mu_j] \geq 0$$

(ii) From Proposition 1 we know that all sellers in CETE demand zero amount of abatement $\tilde{r}_j = 0, \; j = 2,3,\ldots,n$. Their difference in pay off between the CETE and NCEWT can then be written as:

$$[\tilde{r}_N - r_N^{nc}]B_j'(\tilde{r}_N) + \tilde{p}r_j^{nc} - \mu_j + \nu_j \geq 0 \qquad j = 2,3,\ldots,n$$

(iii) Similarly as in *(i)* and *(ii)* we can write the difference in pay off between the CETE and NCEWT for agent *1* as follows:

$$\tilde{p}[r_1^{nc} - r_N^{nc}] - \mu_1 + \nu_1$$

The sign is indetermined because the first term is negative but the two last terms are positive. The latter terms μ_1 and ν_1 are a function of the second derivative of the cost and benefit functions for player 1. Clearly, if both marginal costs and benefits are constant, agent 1 will always lose in CETE compared to NCEWT.

Figure 4.4 illustrates the difference in pay off for the suppliers of abatement between the noncooperative Competitive Emission Trading Equilibrium CETE and the Nash Cournot outcome NCEWT. Country 1 is characterised by the highest marginal benefits from abatement and will become the only purchaser of emission reduction. Country 2 has lower marginal benefits and costs and will sell emission reduction to country 1. The supplier enjoys an increase in benefits because overall abatement is higher with emission trading than without (cfr. the area under country 2's marginal benefit curve between r_N^{nc} and r_N^{ct}). Compared to the Nash Cournot Equilibrium Without Trade, country 2 produces more abatement and incurs a higher cost but this is overcompensated because he gets a price p for every unit of abatement he produces. The net result is always positive so that the supplier wins under the trading regime. The analytical expression in the proof of Proposition 3, section *(ii)*, can easily be identified in Figure 4.4. For instance, if the marginal benefit and costs were constant, the difference in pay off equals $pR_2^{nc} + B'_2(r_N^{ct})[R_N^{ct} - R_N^{nc}]$.

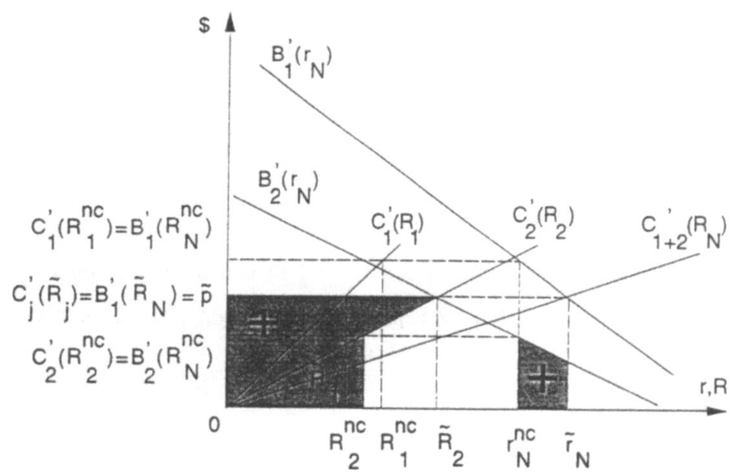

Figure 4.4: Pay off difference for suppliers, CETE versus NCEWT

Figure 4.5 illustrates the case of the emission reduction purchaser, country 1. At the one hand, he enjoys an increase in pay off because his benefits are higher and his costs are lower in CETE than in NCEWT. On the other hand however, he has to pay a price p for every unit of emission reduction he is demanding over his domestic production of abatement. This results in the negative area $-p[r_N^{ct} - R_1^{ct}]$. Whether country 1 wins or loses compared to NCEWT depends on the relative magnitude of the positive and negative effect. Clearly, if marginal cost and benefits were constant, the positive area would vanish and country 1 would be worse off under the emission trading scheme than in the laissez faire outcome. This graphical analysis illustrates the analytical results in the proof of Proposition 3, section *(iii)*.

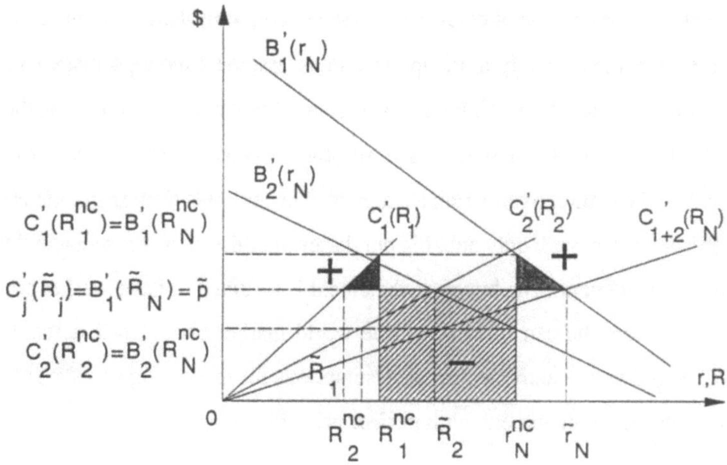

Figure 4.5: Pay off difference for purchaser, CETE versus NCEWT

Two remarks are still to be made. First, the fact that the Competitive Emission Trading Equilibrium does not always satisfy the Voluntary Participation constraint can be used to explain why such a broad scale emission trading scheme has not been observed in reality yet. Given the huge uncertainty surrounding environmental damage estimates and hence also benefit of abatement estimates, it is not surprising that countries are reluctant to enter into a trading mechanism in which they might end up paying the entire bill of emission abatement. If benefits of emission reduction can be estimated more accurately, countries with high willingness to pay for environmental quality are more likely to take initiatives to promote international cooperation on that particular pollution control issue. Secondly, the fact that for some configurations of marginal cost and benefit functions, the unique purchaser of abatement is worse off compared to the Nash Cournot Equilibrium Without Trade does not imply that he is bound to lose. In fact, the question whether a competitive emission trading emission satisfies Voluntary Participation becomes entirely an empirical question. We will take up this question for the world greenhouse problem in section 4.6 of this paper.

4.5 Noncooperative solution with monopsonistic emission abatement trade

Section 4.4 ended with a pessimistic note because competitive abatement trading does not necessarily satisfy Voluntary Participation. However, if there is only one buyer of emission reduction efforts, the assumption of competitive trade in emission abatement might not be very realistic. In this case it is probably more appropriate to assume a monopsonistic structure for the market of emission reduction. Without loss of generality we take country *1* as the only net buyer of emission abatement and assume that all other countries $j=2,\ldots,n$ are net suppliers and price takers on the market of emission abatement. The emission trading market is modelled as a monopsonistic market with only one, big purchaser of emission abatement and with many, small suppliers of abatement. The buyer of abatement is assumed to have market power, i.e., by lowering his demand he can force down the world market price for abatement and vice versa. We will examine two extreme pricing assumptions for this monopsonistic settings: one without price discrimination and one with perfect price discrimination.

Noncooperative monopsonistic emission abatement trading without price discrimination

Under the monopsonistic trading assumption market equilibrium can be written as follows:

$$(R_1 - r_N) + \sum_{j=2}^{n} R_j(p) \equiv (R_1 - R_N) + R_{-1}(p) = 0 \tag{4.11}$$

Total demand for abatement should equal the supply of emission reduction by the monopsonist (agent 1) and by all other suppliers in the market. By restraining its demand for emission abatement, player *1* can lower the price of emission abatement p. Let us define $P(R_{-1})$ as the inverse of the net supply function $R_{-1}(p)$:

$$p = P(R_{-1}) \quad \text{and} \quad P'(R_{-1}) = -\frac{\partial P(.)}{\partial R_1} = \frac{\partial P(.)}{\partial r_N} > 0 \tag{4.12}$$

The sign of the partial derivatives follows from the assumptions on the cost function. The function $P(R_{-1})$ now replaces the fixed price of emission abatement in expression (4.7). This changes nothing for the sellers of abatement, they still consider the price of abatement as fixed and their demand and supply of abatement is still governed by first order conditions (4.8) and (4.9). The unique buyer on the other hand will take into account his effect on the price when determining his optimal supply and demand of abatement:

$$C_1'(R_1) = P(R_{-1}) + P'(R_{-1})[R_1 - r_1] = B_1'(r_N) \tag{4.13}$$

or alternatively, with ϵ_{-1} the price elasticity of the supply of abatement by countries $i = 2, 3, \ldots, n$,

$$C_1'(R_1) = P(R_{-1})\left[1 + \frac{1}{\epsilon_{-1}}\right] = B_1'(r_N) \tag{4.14}$$

The equality on the left hand states that country *1* will increase its effort such that its marginal cost of emission reduction supply is higher than the market price paid by country *1* for all emission reduction achieved by the other countries. The equality on the right hand side implies that country *1* will lower its purchases of emission abatement in order to limit its overall emission reduction budget. By increasing its supply and decreasing its demand for emission reduction country *1* can lower the world price $P(R_{-1}.)$. The extent to which an abatement purchaser can exploit its monopsony power depends upon the price elasticity of supply by the other producers[8].

Given the new first order condition for the monopsonist, we can define the noncooperative Monopsonistic Emission Trading Equilibrium METE.

Equilibrium on the monopsonistic market for emission abatement

Definition 2: *A noncooperative Monopsonistic Emission Trading Equilibrium (METE) consists of a vector of abatement supply* $\tilde{R} \in \times_{j \in N}[0, E_j^0]$, *a vector of emission reduction demand* $\tilde{r} \in \times_{j \in N}[0, E_N^0]$, *and a unique price* $\tilde{p} \in \Re_{++}$ *such that:*

(i) $C'_j(\tilde{R}_j) = \tilde{p} \qquad j = 2, 3, \ldots, n$

(ii) $C'_1(\tilde{R}_1) = \tilde{p}\left[1 + \frac{1}{\epsilon_{-1}}\right]$

(iii) $(\forall i \in N)(\forall r_i \in [0, E_N^0]):\ u_i(\tilde{R}_i, \tilde{r}_i, \tilde{r}_{-i}) \geq u_i(\tilde{R}_i, r_i, \tilde{r}_{-i})$

(iv) $\sum_{j \in N} \tilde{R}_j = \sum_{j \in N} \tilde{r}_j$

Compared to Definition 1, only the first order condition for the monopsonist changes. Under the assumption that the monopsonist cannot discriminate between the suppliers by paying different prices, the monopsonist has to pay the same price $\tilde{p}$ for every single unit of emission

---

[8] This price elasticity of abatement supply by all other producers than the monopsonist can be shown to be (from first order conditions (4.9)):

$$\epsilon_{-1} = \frac{p}{R_{-1}(p)}\frac{\partial R_{-1}(p)}{\partial p} = \frac{p}{R_{-1}(p)}\sum_{j=2,3,\ldots,n}\frac{1}{C_j''(.)} \geq 0$$

If the small abatement suppliers have relatively flat marginal cost functions, the elasticity of abatement supply tends to be higher such that the monopsonist has less market power.

supply no matter who produced it. This implies that the suppliers of emission reduction are able to keep their producer's surplus as in the CETE scenario. Given this pricing assumption, the results of the previous section carry over. For Proposition 4 and 5 we simply report the results but we will not elaborate on the proof. The proofs run similarly to the proof of Proposition 2 and 3.

Proposition 4: *(i) Total emission reduction is higher under the noncooperative Monopsonistic Emission Trading Equilibrium METE without price discrimination than in the corresponding Nash Cournot Equilibrium Without Trade NCEWT.*

(ii) The unique buyer of emission abatement however supplies less emission reduction than in NCEWT.

Proposition 5: *(i) Total utility under noncooperative METE without price discrimination is strictly higher than in the corresponding NCEWT equilibrium.*

(ii) All sellers of emission abatement in the METE without price discrimination are strictly better off than in NCEWT.

(iii) The unique purchaser of abatement might win or lose depending upon the slopes of its marginal benefit and cost functions.

Noncooperative monopsonistic emission abatement trading with perfect price discrimination

It is clear that the pricing assumption in the previous section is rather extreme. Therefore, we assume in the sequel of the paper that the monopsonist can bargain with each supplier separately over the price to pay for its abatement effort. We go even a step further by assuming that the monopsonist has the power to use perfect price discrimination against the producers. This means that he can just compensate the suppliers of emission reduction for their cost of abatement, or equivalently, that the monopsonist can reap all producer surpluses. The suppliers of abatement are paid according to their actual marginal cost of abatement instead of at a fixed price $\bar{p}$ per unit of abatement. It turns out that Propositions 4 and 5 carry over to this pricing assumption as well. The proof of 6 is analogue to the proof of Proposition 2.

Proposition 6: *(i) Total emission reduction is higher under the noncooperative Monopsonistic Emission Trading Equilibrium METE with perfect price discrimination than in the* *corresponding Nash Cournot Equilibrium Without Trade NCEWT.*

(ii) The unique buyer of emission abatement however supplies less emission reduction than in NCEWT.

Proposition 7: *(i) Total utility under noncooperative METE with perfect price discrimination is strictly higher than in the corresponding NCEWT equilibrium.*

(ii) All sellers of emission abatement in the METE with perfect price discrimination are strictly better off than in NCEWT.

(iii) The unique purchaser of abatement might win or lose depending upon the slopes of its marginal benefit and cost functions.

Proof: *(i)* The proof runs similar to the proof of claim (i) in Proposition 3.

(ii) The difference in pay off between the noncooperative METE with perfect price discrimination and NCEWT for the suppliers of abatement are given by:

$$B_j(\tilde{r}_N) - B_j(\tilde{r}_N) + C_j(R_j^{nc}) \geq 0 \qquad j=2,3,\ldots,n$$

(iii) The pay off difference for the monopsonist reads:

$$B_1(\tilde{r}_N) - \sum_{j=2}^{n} C_j(\tilde{r}_N) - B_1(r_N^{nc}) + C_1(R_1^{nc})$$

It is not clear whether the increase in benefits will always outweigh the total compensation paid to the suppliers.

The claims *(ii)* and *(iii)* in Proposition 7 are illustrated in Figure 4.6 and Figure 4.7 respectively.

In order to get some idea of the effects on abatement efforts and for the welfare implications of the monopsonistic market assumption with and without price discrimination, we turn to some simulation exercises in the next section.

4.6 Empirical illustration

Description of the regions and calibration of costs and benefits

We illustrate both the competitive and monopsonistic emission trading equilibrium for a 12 region simulation model for the world carbon economy. The functional form of cost of carbon emission abatement functions in this model are essentially taken from Nordhaus (1991). In his

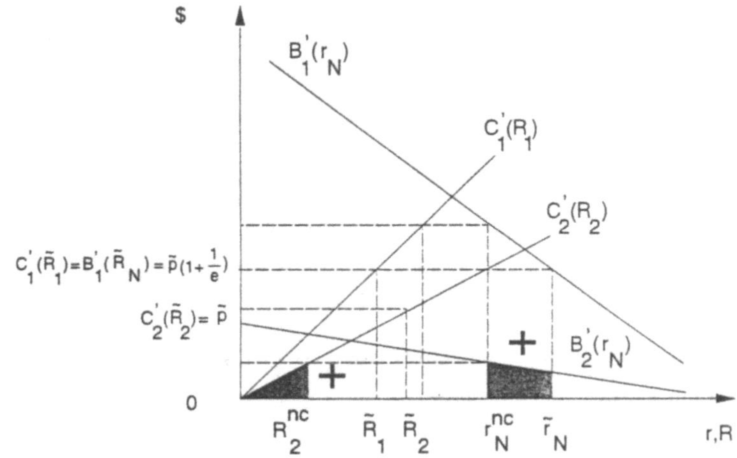

Figure 4.6: Pay off difference for suppliers, noncooperative METE with perfect price discrimination versus NCEWT

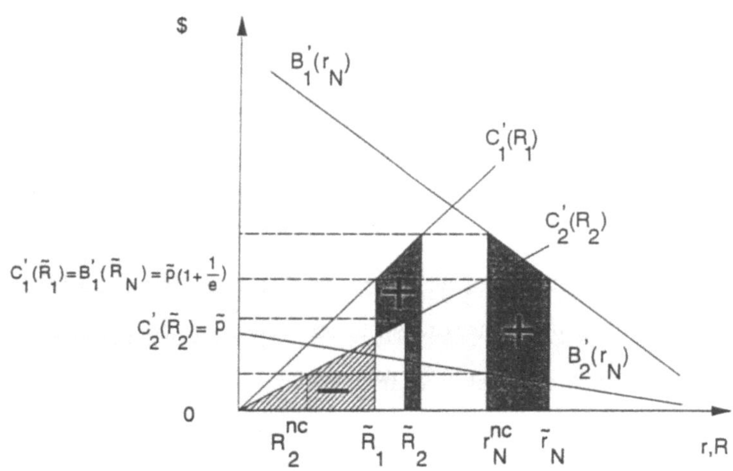

Figure 4.7: Pay off difference for purchaser, noncooperative METE with perfect price discrimination versus NCEWT

formulation marginal costs are a logarithmic function of relative emission reduction:

$$C'_i(R_i) = \frac{1}{\gamma_i} \ln\left(1 - \frac{R_i}{E_i^0}\right) \qquad \gamma_i < 0 \qquad (\forall i \in N) \tag{4.15}$$

We divided the world into 12 regions and for each of these a cost function was calibrated using estimates from the OECD GREEN model in Burniaux e.a. (1992). The advantage of the GREEN results is their regional disaggregation and incorporation of general equilibrium effects. In order to calibrate the parameter γ_i in (4.15) we selected GREEN marginal cost estimates for the year 2010 for a Toronto-type of agreement in which the OECD countries and former USSR reduce their carbon emissions by 20% compared to 1990. This simulation exercise is considered by Burniaux e.a. (1992) themselves as the central case. For some regions, GREEN estimates were not available and we disaggregated the aggregate γ_i parameter in function of the energy intensity of production, i.e., carbon emissions over GDP, of the particular regions[9].

Instead of the constant marginal benefit of emission reduction assumption in Eyckmans e.a. (1993) we opted for a slightly more general quadratic specification of benefits[10]:

$$B_i(r_N) = w_i \, \alpha \left[E^0 - \frac{r_N}{2} \right]^2 \qquad w_i, \ \alpha > 0 \qquad (\forall i \in N) \tag{4.16}$$

As a rough approximation, we assumed benefits to be proportional to GDP. The weight w_i equals the share of country i in 1990 world GDP and shares sum to unity $\sum_{j \in N} w_j = 1$. The parameter α is the slope of the aggregate marginal benefit of abatement function and is calibrated such that the first best optimal carbon tax for the world as a whole equals $\sum_{j \in N} B'_j(R_N^*) = 35.00 US\$ = C'_i(R_i^*) \quad (\forall i \in N)$. This figure was taken from Ayres and Walter (1991) and is relatively high compared to the estimates in, e.g., Nordhaus (1991). The corresponding socially optimal amount of total emission reduction amounts to $R_N^* = 16.25\%$ for the marginal cost estimates described higher. This figure is quite well in line with results in other studies, see, e.g., Fankhauser (1995). Table 4.1 describes some relevant characteristics of the 12 regions in the model as well as the calibrated cost function parameter γ_i.

[9] The exact specification and calibration of the cost functions was described in Eyckmans, Proost and Schokkaert (1993). All data and GAMS simulation programs used to obtain the figures in Table 4.1 and 4.2 are available from the authors upon simple request.

[10] The reason for using quadratic instead of linear benefit functions was simply to obtain well defined emission reduction demand functions.

Table 4.1: Description of the regions

	% emiss.	% GDP	% pop.	GDP capita	emiss. capita	emiss. / GDP	γ_i
North America	26.45	23.10	5.32	20391	5.68	0.28	-0.0029
European Comm.	14.59	17.83	6.59	12698	2.53	0.20	-0.0026
China	11.07	12.04	21.44	2636	0.59	0.22	-0.0194
Pacific	6.63	8.65	2.86	14200	2.64	0.19	-0.0008
ex USSR	16.35	8.11	5.53	6876	3.37	0.49	-0.0069
Latin America	4.72	8.05	8.35	4523	0.64	0.14	-0.0027
Developing Asia	3.37	5.33	7.47	3350	0.52	0.15	-0.0029
Rest Asia	4.27	5.24	23.84	1031	0.21	0.20	-0.0038
Africa	2.99	3.73	12.27	1428	0.28	0.20	-0.0037
Other Europe	1.85	2.93	1.71	8046	1.23	0.15	-0.0020
Eastern Europe	4.75	2.68	2.33	5412	2.33	0.43	-0.0060
Middle East	2.98	2.32	2.31	4721	1.47	0.31	-0.0060
World	100.00	100.00	100.00	4694	1.14	0.24	

North America = US and Canada
European Comm. = European Community (including former Eastern Germany)
Pacific = Japan, Australia, New Zealand and Oceania
ex USSR = all nation of former USSR
Latin America = all countries of American continent excluding N-America
Developing Asia = Hong-Kong, Indonesia, South Korea, Malaysia, Philippines, Singapore, Taiwan and Thailand
Rest Asia = other Asian countries.
Other Europe = non-EC members plus Turkey
Eastern Europe = former East European countries except for ex-USSR and former East Germany

Using the functional forms and parameter values described above, we present simulations for the Nash Cournot Equilibrium Without Trade NCEWT and noncooperative Competitive Emission Trading Equilibrium CETE in Table 4.1. Table 4.2 covers the case of noncooperative monopsonistic emission trading METE with and without price discrimination.

Noncooperative Competitive Emission Trading Equilibrium CETE

The regions in Table 4.2 are ranked according to their marginal benefits of abatement, or according to their share w_i in world GDP. Column (1) to (3) report on the Nash Cournot Equilibrium Without Trade NCEWT. The emission reduction vector r^{nc} was computed to satisfy first order conditions (4.4), i.e., marginal benefits equal marginal costs for every region. Marginal abatement costs in column (2) range between 1.21 US$ for Middle East and 4.50 US$ for North America illustrating the productive inefficiency of the laissez faire equilibrium.

Table 4.2: Nash Cournot Equilibrium Without Trade NCEWT and noncooperative Competitive Emission Trade Equilibrium CETE

	NCEWT			CETE				
	(1)	(2)	(3)	(4)	(5)	(6)	(7)	(8)
North America	2.75	9.41	4.50	2.68	18.19	9.19	9.19	0.94
European Comm.	1.90	7.26	3.08	2.40	0.00	9.19	7.09	6.33
China	9.07	4.90	0.55	16.32	0.00	9.19	4.79	1.71
Pacific	0.27	3.52	3.63	0.71	0.00	9.19	3.44	6.86
ex USSR	2.25	3.30	1.64	6.13	0.00	9.19	3.22	4.24
Latin America	0.87	3.28	1.15	2.43	0.00	9.19	3.20	2.23
developing Asia	0.62	2.17	0.86	2.61	0.00	9.19	2.12	1.66
rest Asia	0.82	2.13	0.26	3.47	0.00	9.19	2.08	0.52
Africa	0.55	1.52	0.37	3.30	0.00	9.19	1.49	0.72
other Europe	0.24	1.19	2.06	1.80	0.00	9.19	1.17	3.94
eastern Europe	0.66	1.09	1.38	5.39	0.00	9.19	1.07	3.16
Middle East	0.56	0.94	1.21	5.32	0.00	9.19	0.92	2.61
world	2.56		1.12	4.81	4.81			1.99

The unique price for emission abatement in CETE amounts to $\tilde{p} = 9.19 US\$$ per tonne of carbon

(1) percentage emission abatement in NCEWT r_i^{nc} / E_i^0
(2) marginal abatement cost in NCEWT in US$ per tonne of carbon
(3) per capita pay off in NCEWT in US$
(4) percentage emission abatement supply in CETE $\tilde{R}_i / E_i^0$
(5) percentage emission abatement demand in CETE $\tilde{r}_i / E_i^0$
(6) marginal abatement cost in CETE in US$ per tonne of carbon
(7) marginal abatement benefit in CETE in US$ per tonne of carbon
(8) per capita pay off in CETE in US$

Lets consider the noncooperative Competitive Emission Trading Equilibrium CETE in column (4) to (8). The emission reduction supply vector $\tilde{R}$ in column (4) and demand vector $\tilde{r}$ in column (5) were computed by means of the first order conditions (4.8) and (4.9) respectively. Turning to the demand for abatement, we notice that North America, which is characterized by the largest marginal benefits, is the only region that demands a strictly positive amount of emission abatement. All other regions are selling their surplus of abatement to North America. Notice that in particular China chooses to produce a surplus of abatement because it is characterized by the lowest marginal cost for carbon emission reduction[11]. In general we notice

[11] For the cost functions used in this simulation, one could make a case that China has some monopoly power in the supply of abatement since it is by far the cheapest producer. The analysis of section (4.5) can be translated straightforward to a situation were only one supplier of abatement exercises monopoly power.

that total emission abatement, $\tilde{R}_N = 4.81\%$ is higher if we allow for emission reduction trading compared to Nash Cournot Without Trade abatement $R_N^{nc} = 2.56\%$ but it stays far below the first best optimum of about 16% of abatement. The trading mechanism does not achieve a spectacular increase in the total provision of the public good but at least abatement is produced in a cost efficient way as can be seen from the marginal cost figures in column (6). For every region, marginal abatement costs equal the price of abatement $\tilde{p} = 9.19US\$$. Column (7) illustrates Proposition 1 of section 4.4. Only North America can equalize its marginal abatement costs to its marginal benefit. All suppliers of abatement are constrained to marginal benefits strictly below the market price for abatement. Compared to the NCEWT, total utility is higher in CETE, 1.99 versus 1.12 US$ per capita. All suppliers of emission reduction are strictly better off in terms of utility in CETE than in NCEWT. The unique purchaser of abatement, North America, suffers an important decrease in pay off, 0.94 US$ per capita in CETE against 4.50 US$ per capita in NCEWT. Obviously, the proposal would never be accepted by North America.

Noncooperative Monopsonistic Emission Trading Equilibrium METE

For the noncooperative Monopsonistic Emission Trading Equilibrium, we assumed that all regions except North America are taking the price of emission abatement as given. For North America we corrected the first order conditions by the appropriate derivative of the inverse supply of abatement curve of the abatement producers other than the monopsonist, cfr. first order conditions in Definition 2. For these behavioural assumptions we again calculated a vector of emission supply and emission demand and a new price level such that the emission market clears. Pay off figures in column (5) in Table 4.3 assume that North America cannot discriminate against the suppliers of abatement by charging individualized prices. In column (6) however, North America reaps all producers benefits since it pays every seller of emission abatement according to its specific marginal abatement cost. The latter case can be interpreted as North America running some kind of greenhouse fund and letting the suppliers bit against each other to produce abatement.

Table 4.3: Noncooperative Monopsonistic Emission Trade Equilibrium METE

	(1)	(2)	(3)	(4)	(5)	(6)
North America	2.68	18.10	9.19	9.19	0.99	5.06
European Comm.	2.38	0.00	9.13	7.09	6.30	6.06
China	16.22	0.00	9.13	4.79	1.70	1.26
Pacific	0.71	0.00	9.13	3.44	6.82	6.77
ex USSR	6.09	0.00	9.13	3.23	4.21	3.28
Latin America	2.42	0.00	9.13	3.20	2.22	2.16
developing Asia	2.60	0.00	9.13	2.12	1.65	1.59
rest Asia	3.44	0.00	9.13	2.08	0.52	0.49
Africa	3.28	0.00	9.13	1.49	0.72	0.68
other Europe	1.79	0.00	9.13	1.17	3.92	3.84
eastern Europe	5.36	0.00	9.13	1.07	3.14	2.58
Middle East	5.29	0.00	9.13	0.92	2.60	2.25
world	4.79	4.79			1.98	1.99

The unique price for emission abatement in METE amounts to $\tilde{p} = 9.13 US\$$ per tonne of carbon. The price elasticity of abatement supply of price taking producers equals $\epsilon_{-1} = 0.61$.
(1) percentage emission abatement supply in METE $\tilde{R}_i / E_i^0$
(2) percentage emission abatement demand in METE $\tilde{r}_i / E_i^0$
(3) marginal abatement cost in METE in US$ per tonne of carbon
(4) marginal abatement benefit in METE in US$ per tonne of carbon
(5) per capita pay off in METE without price discrimination in US$
(6) per capita pay off in METE with price discrimination in US$

Compared to the competitive trade case CETE, the price of abatement is somewhat lower in the monopsony scenario METE: 9.13 US$ versus 9.19 US$ per tonne of carbon because the North America lowers its demand from 18.19% in CETE to 18.10% in METE. At that price all suppliers produce slightly less abatement because they equalize their marginal cost to the market price for abatement as can be seen from column (3). The monopsonist on the other hand produces emission reduction up to the point where $C'_1(\tilde{R}_1) = \tilde{p}[1 + 1/\epsilon_{-1}]$. He ends up on approximately the same level of abatement supply as in the CETE case. As a result, total emission reduction is only 0.02% lower in the monopsony scenario compared to the competitive trade scenario. If price discrimination is impossible North America achieves a slightly higher pay off in METE than in CETE but still, this figure stays far below North America's pay off in the laissez faire equilibrium, 0.99 US$ versus 4.50 US$. All other regions are losing a small fraction of their pay off compared to CETE but they are still strictly better of than in NCEWT. If on the other hand price discrimination by North America is possible, all sellers of abatement lose their producer surplus and this is not compensated for under the form of higher total emission reduction. North America is better off if it can discriminate against the suppliers of abatement because it can reap the total amount of producers surpluses. If we compare the NCEWT with the METE scenario with price discrimination, it turns out that both North America and the suppliers of abatement achieves a higher pay off than in the Nash Cournot Equilibrium Without Trade NCEWT. Hence, the METE with price discrimination satisfies the Voluntary Participation constraint and constitutes a Pareto improvement over the laissez faire Nash Cournot Equilibrium Without Trade.

However, as was shown in the previous section, this result does not hold in general. It is possible to construct emission reduction games where the suppliers of emission are strictly better off than in NCEWT but where the monopsonist loses even if he can charge individualized prices.

4.7 Conclusion

In this paper we examine how Pareto improvements starting from the noncooperative Nash Cournot Equilibrium Without Trade can be achieved by introducing possibilities for emission trading. We suggested to set up an international market where individual countries can trade emission reduction. This is not the same as a traditional emission permits market because we do not distribute emission entitlements at the start. In our trading mechanism all countries determine their optimal supply and demand of abatement and buy or sell surpluses at a fixed price on the abatement market. Under the assumption that trade in abatement is competitive, it was shown that the market system does not always satisfy Voluntary Participation for all possible configurations of cost and benefit of abatement functions. For quasi linear preferences, it turns out that there is only one purchaser of abatement who might end up worse off after trading took place than in the Nash Cournot Equilibrium Without Trade, depending upon the relative slopes of marginal cost and benefit functions. In order to allow for more flexibility, we relaxed the assumption of price taking behaviour on the emission abatement market to allow for monopsonistic behaviour on part of the purchaser. Even if the purchaser was given the opportunity to apply perfect price discrimination against the suppliers of abatement, it is still possible that the purchaser loses compared to its reservation pay off level in the Nash Cournot Equilibrium Without Trade.As the theoretical analysis could not give definite answers to the question whether an Emission Trading mechanism could satisfy the Voluntary Participation constraint, we turned to some simulation exercise for the world greenhouse problem. Simulations for simple specification of the world carbon economy indicate that one has to accepts that the monopsonist, i.e. North America, reaps part of the producers' surplus by means of price discrimination. Otherwise, also monopsonistic trading would violate Voluntary Participation in this simulation exercise. Overall, carbon emission reduction increased from some 2.6% in Nash Cournot NCEWT to about 4.79% if we allow for emission abatement trade. This is a small improvement compared to the first best level of 16.25% abatement. However, IEA (1992) estimated that if all unilateral (and voluntary) abatement commitment made after the 1992 Rio Earth Summit were to be effectively realized, emissions of carbon dioxide would be reduced by slightly less than 3% by the year 2000. Our simple simulations for noncooperative and voluntary emission abatement trading yield a figure in the same order of magnitude.

Finally, we would like to suggest two directions for further research. First, the assumption of quasi linear preferences should be relaxes to allow for income effects. Proposition 1, that there would be only one agent demanding a strictly positive amount of abatement, is unlikely to carry over because marginal willingness to pay for total emission reduction can adapt to equal the international price of abatement due to income effects. Secondly, given the large share of one producer, China, in the supply of abatement, it is highly probably that this country will try to exploit its monopoly power. A similar analysis as for the monopsonistic trading assumption should be developed.

References

Ayres, R.U., and Walter, J. (1991) The greenhouse effect: damages, costs and abatement. *Environmental and Resource Economics, 1, pp. 237-270.*

Barrett, S. (1992a) *Conventions on climate change*. OECD, Paris.

Barrett, S. (1992b) Transfers and the gains from trading carbon emission entitlements in a global warming treaty. In: UNCTAD (Ed.). *Combating global warming, study on a global system of tradable carbon emission entitlements*, UNCTAD - UN, New York.

Burniaux, J.M., Martin, J.P., Nicoletti, G., and J. Oliveira-Martins (1992) *The costs of reducing CO2 emissions, evidence from GREEN*. OECD Economics Department, Working Paper, nr. 115, OECD, Paris.

Coase, R. (1960) The problem of social cost. *Journal of Law and Economics*.

Eyckmans, J., Proost, S. and E. Schokkaert (1993) Efficiency and distribution in greenhouse negotiations. *Kyklos 46, pp. 363-397.*

Eyckmans, J., Proost, S. and E. Schokkaert (1994) A comparison of three international agreements on carbon emission abatement. In: E. van Ierland (Ed.) *International environmental economics - theories, models and applications to climate change, international trade and acidification*, Elsevier Science, Amsterdam.

Fankhauser, S. (1995) *Valuing Climate Change*. Earthscan, London.

Hahn, R.W. (1984) Market power and transferable property rights. *The Quarterly Journal of Economics, X, pp. 753-765.*

Hoel, M. (1992a) International environment conventions: the case of uniform reductions of emissions. *Environmental and Resource Economics, 2, pp. 141-159.*

Hoel, M. (1992b) Carbon taxes - an international tax or harmonized domestic taxes? *European Economic Review, 36, pp. 400-406.*

International Energy Agency (1992) *Climate Change Policy Initiatives*. IEA, OECD, Paris.

Nordhaus, W.D. (1991) The cost of slowing climate change: a survey. *The Energy Journal, 12, pp. 37-65.*

5 On the Efficiency of Green Tax Reforms to Reduce CO_2 Emissions*

Ronnie Schöb
Lehrstuhl für Nationalökonomie und Finanzwissenschaft
University of Munich
Schackstr. 4
D-80539 Munich, Germany

Abstract

According to the double-dividend hypothesis, the introduction of carbon taxes is expected to both reduce carbon dioxide emissions and the distortions of the existing tax system. Contrary to the common belief that emissions will decrease, this paper points out that an increase of carbon dioxide emissions due to the introduction of carbon taxes within a green tax reform cannot be ruled out in general. To determine the effect of such a green tax reform on the environment one has to take account of the possible feedback effects of all accompanying measures taken by a government. In the case of a revenue-neutral green tax reform, these may be tax rate cuts for at least one non-polluting good. Simulations show that an increase of carbon dioxide emissions due to a revenue-neutral introduction of carbon taxes might occur in an empirically relevant range of parameters.

---

* Comments by two anonymous referees are gratefully acknowledged. The usual disclamer applies.

NATO ASI Series, Partnership Sub-Series, 2. Environment – Vol. 14
Economics of Atmospheric Pollution
Edited by Ekko C. van Ierland and Kazimierz Górka

5.1 Introduction

Atmospheric pollution has become a global problem which calls for global solutions. According to the reports of the Intergovernmental Panel on Climate Change (Houghton *et al.*, 1990, 1992), the rapidly growing emissions of greenhouse gases, in particular the emissions of carbon dioxide (CO_2) will lead to an increase in average global temperature of between 1.5°C and 4.5°C. The sea-level is expected to rise up to 70 cm within the next century. Of course, there is great uncertainty about the consequences of the climate change and the sea level rise with regard to health and mortality, agriculture and forestry in the future. Therefore the external cost the present generation is imposing on future generations is also highly uncertain (cf. Cline, 1992). Nevertheless, there is wide agreement that the net effect will be negative and that at least some global environmental policy measures are required to reduce CO_2 emissions.

Economists in general believe in the efficiency of taxes on emissions. It is therefore no longer surprising that proposals to introduce an international carbon tax to reduce carbon dioxide emissions are on the agenda of almost all international conferences on global warming. More recently, such tax reform proposals have received even more support from the double dividend argument. The double-dividend hypothesis claims that green taxes do not only improve the environment but allow the government to reduce the distortions of the existing tax system by cutting other taxes (cf. Pearce, 1991).

The double-dividend literature, however, focuses on the effect a green tax such as the carbon tax has on the efficiency of the tax system (cf. Repetto and Dower, 1992; Bovenberg and de Mooij, 1994). Thereby, it has widely ignored the impact the changes within the tax system will have on the environment (one rare exception is Ng, 1980, see section 4 below). It seems to be taken for granted that carbon dioxide emissions will be reduced by introducing a carbon tax (cf. e.g. Goulder, 1995). Such claims, however, focus on the carbon tax alone and disregard the effects of the accompanying government measures on the environment.

The analysis of second-best policies emphasises the need to take account of all possible feedback effects of the accompanying measures taken by a government. In the case of introducing a carbon tax within a revenue-neutral green tax reform, accompanying measures comprise tax rate cuts for non-polluting (clean) goods. These tax rate cuts for clean goods might have some impact on emissions because of the complementarity/substitutability relationships which might exist between clean goods and polluting goods.

The paper tries to close this gap in the literature by analysing the impact of both a carbon tax and the accompanying tax rate cuts on the environment. Although it follows the public economics literature on tax reform analysis (e.g. Guesnerie, 1977; Ahmad and Stern, 1984), the main purpose of the paper is not to identify welfare improving tax reforms but to analyse the overall impact of green tax reforms on CO_2 emissions.

It will be shown that in a first-best world the introduction of a carbon tax, accompanied by a lump-sum rebate of tax revenues, always improves the environment (section 3). In a second-best framework, however, a rebate of the carbon tax revenues via the reduction of some other commodity taxes may, under certain circumstances, actually cause the environment to deteriorate. As we cannot rule out the paradoxical case, this paper derives conditions which guarantee that the environmental impact of green tax reforms becomes positive (section 4). Section 5 then investigates the empirical relevance of these effects. By providing some simulations, it will be shown that there is some empirical evidence that the paradoxical case may actually occur. Section 6 concludes.

5.2 The model

We consider the decision problem of a social planner (e.g. a benevolent supranational government) of the present generation who has to deal with the problem of an intertemporal externality. This will be done in a simple two-period framework.

The present generation

Ignoring distributional issues, we assume that in both periods there are H identical households. Each household consumes the private goods 0, c, d. The preferences of the representative household of the present period p are described by a twice continuously differentiable, strictly quasi-concave utility function:

$$u^p(x_0, x_c, x_d).$$

The marginal utilities are positive, u_i, $i = 0, c, d$. The production of the private goods takes place competitively, subject to a linear technology. All producer prices are thus constant and are normalised at unity. Consumer prices therefore vary only with varying commodity taxes.

Without loss of generality, we can normalise the tax on good 0 at zero, $t_0 = 0$. The household's budget constraint is then given by:

$$x_0 + (1+t_c^p)x_c + (1+t_d^p)x_d = T^p.$$

The tax rate t_i^p indicates the present tax rate on good i. T^p is the present lump-sum transfer from the government to the household. The household's initial endowments are normalised to zero for all goods.

Using the indirect utility function v, the welfare of the present generation can be represented by the utility of the representative household:

$$v^p(t_c^p, t_d^p, T^p). \tag{5.1}$$

Emissions

The present consumption of the good d, called the dirty good, leads to CO_2 emissions and hence contributes to the greenhouse effect from which the future generation suffers. Total emissions E are equal to the aggregate consumption of the dirty good. Using the individual demand functions and assuming a one-to-one relationship between the consumption of the dirty good and emission, the emission function is given by:

$$E = \sum_{h=1}^{H} x_d^h(t_c^p, t_d^p, T^p). \tag{5.2}$$

Neither production nor the consumption of the clean good c do contribute to total emissions of CO_2.

The future generation

Technology and preferences are the same as in the present period. However, each future generation household suffers from the present generation's emission E. Hence, emissions enter the utility function of future households. Using the indirect utility function for the representative future generation household, the future generation's welfare can be represented by:

$$v^f(t_c^f, t_d^f, T^f, E), \tag{5.3}$$

whereby the marginal utility of emissions v_E is negative for all positive amounts of E.

The social planner

In each generation all households are treated equally by the government. The government maximises a welfare function w which considers the utility of the representative present generation household and the utility of the representative future generation household. Abstracting from the problem of discounting, the social welfare function of the Benthamite type is given by:

$$w = v^p(t_c, t_d, T) + v^f(t_c^f, t_d^f, T^f, E). \tag{5.4}$$

The only control variables of the present social planner are the present period's tax rates and lump-sum transfers. When maximising welfare, however, the present social planner can take into account the fact that the future social planner will have maximised the future generation's welfare by choosing an optimal set of future taxes $(t_c^{f*}, t_d^{f*}, T^{f*})$. Note that, in so far as future demand depends on emissions E, the optimal future tax rates will depend on E.[1]

For the present social planner, the future generation's welfare can be described by:

$$v^f(E) = v^f(t_c^{f*}(E), t_d^{f*}(E), T^{f*}(E), E). \tag{5.5}$$

Hence, the objective function reduces to:

$$w = v^p(t_c, t_d, T) + v^f(E). \tag{5.4'}$$

The marginal welfare of emissions E is given by $w_E = v_E^f < 0$. Note that the present generation does not suffer from global warming. In what follows we suppress the superscripts p and f.

Welfare changes

The welfare change of a tax rate change, normalised by the marginal utility of the numéraire u_0 (i.e. the marginal utility of present lump-sum income), is given by:

$$d\tilde{w} \equiv \frac{dw}{u_0} = \left(-x_k + \frac{w_E}{u_0} \frac{\partial E}{\partial t_k} \right) dt_k, \tag{5.6}$$

with $k = c, d$. The first term in brackets denotes the direct utility loss of the present household according to Roy's identity. The second term denotes the indirect effect of a tax rate change due to the change in the environmental quality tomorrow. As emissions do not matter in the present period, the consumption of the taxed good is independent of the environment. Hence,

[1] Hereby, we have ruled out intergenerational transfers other than via E. Otherwise we would have to consider the possibility of compensating the future generation by some non-environmental capital.

we can use $\partial E/\partial t_k = H \cdot \partial x_d/\partial t_k$ in the second term. Because of the own-price effect ($k = d$) or the cross-price effect ($k = c$), a change in the consumer price changes the consumption of the dirty good d and therefore changes the emissions E. The budget constraint for the government (per capita) is given by

$$R = \sum_{k=c,d} t_k x_k - T. \tag{5.7}$$

R indicates a given tax requirement. The first term covers the revenues due to commodity taxation while the second term denotes the lump-sum transfer to the household. (The transfer might also be negative.) Fixing R at a certain level allows us to focus on revenue neutral tax reforms.

5.3 First-best analysis

Assume that, before the carbon tax is introduced, the government does not care about the future generation's environment quality - maybe because the effects of global warming and the consequences are still unknown. In a world without any restrictions on the use of tax instruments, the government finances the public good provision by lump-sum taxes only: $R = -T,\ t_i = 0,\ \forall i$. To internalise the externality (at the margin) the government therefore levies a tax on the dirty good and refunds the revenues via a reduction in lump-sum taxes. Revenue neutrality requires:

$$\mathrm{d}R = x_d \mathrm{d}t_d - \mathrm{d}T = 0. \tag{5.8}$$

The welfare change of such a tax reform is given by

$$\mathrm{d}\tilde{w} = \left(-x_d + \frac{w_E}{u_0} H \frac{\partial x_d}{\partial t_d}\right)\mathrm{d}t_d + \left(1 + \frac{w_E}{u_0} H \frac{\partial x_d}{\partial T}\right)\mathrm{d}T. \tag{5.9}$$

Adding the revenue neutrality condition (5.8) to the welfare change (5.9) and making use of the total derivative of (5.2), we obtain:

$$\mathrm{d}\tilde{w} = \frac{w_E}{u_0}\mathrm{d}E. \tag{5.10}$$

If there are no externalities, a small revenue-neutral variation of prices and income has no effect on welfare. However, in the presence of externalities we have to consider the welfare implications of the external effect. The welfare change is equal to the marginal environmental damage the present generation imposes on the future generation.

In a first-best world the introduction of a carbon tax will always lead to a reduction of emissions. To see this, we have to solve the revenue neutrality condition for $\mathrm{d}t_d$ and apply the Slutsky equation to the total derivative of equation (5.2). The welfare change then becomes

$$\frac{\mathrm{d}\tilde{w}}{\mathrm{d}t_d} = H\frac{w_E}{u_0}\left(\frac{\partial x_d}{\partial t_d} + x_d\frac{\partial x_d}{\partial T}\right) = \frac{w_E}{u_0}Hs_{dd}, \tag{5.11}$$

where $s_{dd} < 0$ denotes the (negative) compensated own-price effect. The marginal welfare of emissions w_E is negative by definition. The right-hand side is therefore always positive: starting from the initial situation without taxes - except lump-sum taxes - the first-best tax reform will always reduce emissions and therefore improve welfare.[2]

5.4 Second-best analysis

As long as the government is not restricted in choosing the appropriate instruments, we can proceed in a first-best framework. Normally, however, the government does face some restrictions. In particular, the government is not allowed to use any lump-sum taxes or transfers. Instead of lump-sum transfers, the government is forced to pay back the tax revenues from increasing the tax on the dirty good by subsidising the clean good. In this case, for $T = 0$ and $t_i \geq 0, \forall i$, the revenue neutrality condition is given by[3]

$$\mathrm{d}R = \left(x_d + \sum_{i=c,d} t_i\frac{\partial x_i}{\partial t_d}\right)\mathrm{d}t_d + \left(x_c + \sum_{i=c,d} t_i\frac{\partial x_i}{\partial t_c}\right)\mathrm{d}t_c = 0 \equiv MR_d\mathrm{d}t_d + MR_c\mathrm{d}t_c. \tag{5.12}$$

---

[2] If we continue to increase the tax on the dirty good, we will end up with an optimal tax which is equal to the marginal environmental damage, $t_d^* = -w_E/u_0$. This result is well known from the environmental economics literature (cf. Baumol and Oates 1988).

[3] The exclusion of lump-sum taxes in identical consumer economies is somehow ad hoc as there are no incentive compatibility or information problems present. Nevertheless this is a common assumption made in the optimal taxation literature to focus on efficiency aspects only. The case where a lump-sum rebate (at least at the margin) is possible in the presence of other distortionary taxes, the so-called 'eco-bonus', is analysed in Schöb (1995). Here the paradoxical case becomes less likely but it cannot be ruled out in general. In what follows we assume positive marginal tax revenues without restricting the validity of the results. In the case of negative marginal revenues, only the sign of the tax rate changes has to be changed.

The welfare change due to this tax reform is

$$\mathrm{d}\tilde{w} = \left(-x_d + \frac{w_E}{u_0} H \frac{\partial x_d}{\partial t_d}\right)\mathrm{d}t_d + \left(-x_c + \frac{w_E}{u_0} H \frac{\partial x_d}{\partial t_c}\right)\mathrm{d}t_c. \tag{5.13}$$

The first term in each bracket of the right-hand side indicates the change in consumer surplus due to the change in the consumer price. As the price change results from a tax rate change, this term describes the change in the efficiency of the tax system. The second term describes the welfare change due to the change in emissions of carbon. The total change in emissions is obtained by total differentiation of equation (5.2):

$$\mathrm{d}E = H \frac{\partial x_d}{\partial t_d}\mathrm{d}t_d + H \frac{\partial x_d}{\partial t_c}\mathrm{d}t_c \ . \tag{5.14}$$

Solving equation (5.12) for $\mathrm{d}t_c$, replacing $\mathrm{d}t_c$ in equations (5.13) and (5.14), and inserting equation (5.14) into equation (5.13), we finally obtain

$$\frac{\mathrm{d}\tilde{w}}{\mathrm{d}t_d} = (-MCF_d + MCF_c) + \frac{w_E}{u_0}\frac{\mathrm{d}E}{\mathrm{d}t_d}. \tag{5.15}$$

We thereby make use of the definition of the marginal cost of public funds $MCF_i = -x_i / MR_i$ (cf. e.g. Mayshar 1990). The interpretation of the first term is straightforward. The consumer surplus, and hence the efficiency of the tax system, increases if the marginal cost of public funds of the good whose tax rate is reduced is higher than the marginal cost of public funds of the taxation of the dirty good. The second term on the right-hand side describes the change of welfare due to the change of CO_2 emissions.

The purpose of this paper is to analyse the second effect. We will therefore neglect the first term and focus on the second. This approach can be justified by assuming that the government has already maximised the efficiency of the tax system while disregarding the environment. Hence, the first term becomes zero. The welfare change is positive (negative) only if emissions decrease (increase). Such a tax system may be described as a Ramsey-optimal tax system.[4] The Ramsey optimum may also be interpreted as an optimal tax policy before the consequences of carbon dioxide emissions on climate change and hence the existence of external costs for the future generation have become known to the present government.

---

[4] For a general treatment of both effects for the evaluation of tax reforms, see Schöb (1994).

Solving equation (5.12) for dt_c and inserting in (5.14), the change of emissions can be described by:

$$\frac{dE}{dt_d} = \frac{\partial x_d}{\partial t_d} - \frac{MR_d}{MR_c}\frac{\partial x_d}{\partial t_c}. \tag{5.16}$$

From equation (16) we can derive the following condition:

$$\frac{dE}{dt_d}\begin{Bmatrix} < \\ = \\ > \end{Bmatrix} 0 \Leftrightarrow \frac{\dfrac{\partial x_d}{\partial t_c}}{\dfrac{\partial x_d}{\partial t_d}} \begin{Bmatrix} < \\ = \\ > \end{Bmatrix} \frac{MR_c}{MR_d}. \tag{5.17}$$

Emissions reduce if, and only if, the ratio of the cross-price effect on the dirty good to the own-price effect of the dirty good is less than the ratio of the marginal tax revenues. However, if the dirty good is a substitute for the clean good, i.e. $\partial x_d / \partial t_c > 0$, the left-hand side will be negative and emissions will be reduced by both increasing the tax on the dirty good and reducing the tax on the substitute.[5]

In the case of a complementary relationship between the two taxed goods, i.e. $\partial x_d / \partial t_c < 0$, the change in emissions is no longer clear. The reform improves the environment if, and only if, the reduction in the consumption of the dirty good due to its own-price increase is higher than the increase due to the price reduction of the complement. Hence, as can be seen from equation (5.15), welfare increases if, and only if, the emissions reduce.

To see why this need not be the case, assume that the government increases the tax on the dirty good by one unit. If the marginal revenue MR_d is very high, the additional funds the government raises are large. These have to be refunded via subsidising the clean good by $dt_c < 0$. If the marginal revenue of the clean good MR_c is relatively low compared to MR_d, the clean good will be subsidised at a high rate. As can be seen from (5.16), the subsidy - determined by MR_d / MR_c - is just the weight of the cross-price effect $\partial x_d / \partial t_c$. If the weight is large compared to the weight of the own-price effect (= 1), it might happen that, even in the case of a low cross price-effect relative to the own-price effect, the increase in the consumption of the dirty good resulting from a reduction in t_c outweighs the reduction in consumption resulting from an increase in t_d. Proposition 1 summarises:

[5] In what follows all complementarity/substitutability relationships are uncompensated.

Proposition 1: In a world with distortionary taxation, the introduction of a carbon tax within a revenue-neutral marginal green tax reform reduces emissions, if and only if (*i*) the accompanying tax reduction applies to a substitute for the dirty good or (*ii*) if it applies to a complement to the dirty good and the ratio of the cross-price effect on the dirty good to the own-price effect of the dirty good is smaller than the ratio of the associated marginal tax revenues.

The proposition is similar to proposition 1 in Schöb (1994) who analyses the validity of the double-dividend hypothesis within a more general tax reform framework. Surprisingly, there has not been much concern about the interrelationships between taxes on emissions and other taxes. Indeed, to the best of my knowledge, there is only one paper by Ng (1980) which deals with this issue. Ng derives similar results (see his proposition 1) by looking at labour taxation. He states that welfare will increase, "provided that an increase of the (consumer) price of the externality-producing good is more effective in reducing its consumption proportionately than is an increase in the (consumer) price of labor in increasing it, proportionately to labor" (Ng, 1980: p. 745).

However, even though he assumes revenue neutrality, he does not recognise that the effectiveness of price changes depends on the marginal tax revenues. (See his equation (5.15) and the following discussion of his results.) Instead, Ng abstracted "from the complication of a positive revenue requirement" (Ng, 1980: p. 747) when interpreting his result. Condition (5.17), however, shows that, because of the revenue-neutrality condition, the marginal tax revenues actually determine the relative magnitudes of the tax rate changes and thus the 'relative effectiveness' of the price changes.

5.5 Some remarks on the relevance

The question arises of whether the result which is summarised in proposition 1 is a mere theoretical curiosity or whether it might be of some empirical relevance for actual tax reform proposals. Figure 1 shows how in the framework of the last section the change in emissions due to a one percent increase of the tax rate on gasoline, which can be interpreted as a carbon tax because of the proportionality between consumption and CO_2 emissions (see equation (5.2)).

To calculate the changes of emissions due to a revenue neutral introduction of a carbon tax, we have to reformulate the revenue-neutrality condition. We therefore use the definition $R_i = t_i x_i$ for the tax revenues due to the commodity taxation of good i and the definition for the tax elasticities,

$$\tau_{ij} = \frac{t_j}{p_j}\varepsilon_{ij} = \frac{t_j}{p_j}\frac{p_j}{x_i}\frac{\partial x_i}{\partial p_j}, \; i, j = c, d, \tag{5.18}$$

whereby e_{ij} denotes the uncompensated price elasticity of good i with respect to the price change of good j. Equation (5.12) then becomes:

$$\text{d}R = \left((1+\tau_{dd})R_d + R_c\tau_{cd}\right)\frac{\text{d}t_d}{t_d} + \left((1+\tau_{cc})R_c + R_d\tau_{dc}\right)\frac{\text{d}t_c}{t_c} = 0. \tag{5.12'}$$

The change in emissions can be written as:

$$\frac{\text{d}E}{E} = \tau_{dd}\frac{\text{d}t_d}{t_d} + \tau_{dc}\frac{\text{d}t_c}{t_c}. \tag{5.14'}$$

As an example, consider the proposal of the German Liberal Party (F.D.P., 1993) to increase gasoline taxes and instead to abolish taxes on cars. The simulation uses the following parameters, whereby the tax rates are roughly related to the German tax system. The tax on gasoline is about 2/3 of the consumer price, the inclusive price of the clean good is assumed to be 0.2 which is equivalent to a 15% value added tax rate on cars plus the particular car tax.[6]

In addition, figure 5.1 shows that this effect may occur in a range of elasticities which are not so unrealistic. The environmental effect of a green tax reform is already negative, if, e.g. the own-price elasticity of the clean good ε_{cc} = -0.9 and the cross price elasticity ε_{dc} is less than –0.175. For $\varepsilon_{cd} = \varepsilon_{dc} = -0.2$ a price elasticity of $\varepsilon_{cc} = -0.3$ for the clean good is sufficient to obtain the paradoxical case.

[6] By relating these tax rates to the German tax system we implictly assume zero taxes on labour income. According to the German Ministery of Finance, 8.5 million out of 30 million households liable to taxation pay no income taxes at all. In the model presented in section 2 we have normalized the labour tax rate to zero. This implies that for progressive income taxes the simulation would have to consider different tax rates for different income groups to obtain the effect on carbon dioxide emissions within the whole economy.

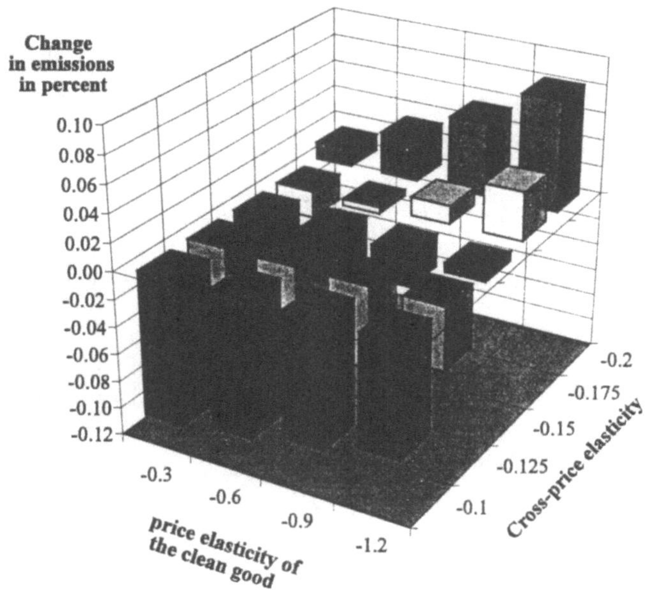

Figure 4.1: The percentage change of emissions due to a one percentage increase of carbon taxes within a revenue neutral tax reform.

It should be emphasised that these results are very sensitive to the ratio of total revenues. In our example, we have assumed that $R_d / R_c = 5$ which is rather high. In Germany, however, there is already a ratio of fuel tax revenues to car tax revenues of 4.3 (adding the VAT to the total car tax revenues, this reduces to 2). Following the proposal of the German Liberal Party (F.D.P., 1993) to increase gasoline taxes and to abolish taxes on cars instead might lead - at least at the margin - into the relevant range.

It is not the main purpose of this paper to find empirical evidence for the paradoxical case. Instead, this example will shed some light on the importance of the accompanying measures taken by the government for the efficiency of green tax reforms. Revenue-neutrality has become an important political argument in support of the introduction of carbon taxes. In a world with distortionary taxation, the efficiency of green tax reforms can be estimated correctly only if we take account of the environmental impact of all accompanying tax rate cuts because of the complementarity/ substitutability relationships between taxed good.

We are accustomed to talking about the efficiency of green taxes. But we should make a careful distinction between the effect a *carbon tax* and the effects a *green tax reform*. Taking the ecological perspective more generally, we can also recognise beneficial incentives of all taxes on goods to which pollution is a complement. Conversely, we also have to be aware of the additional burden of all taxes on goods for which pollution is a substitute.

As condition (5.17) shows, looking for a substitute like public transportation for private car travel can guarantee a double positive effect on the future environment. Disregarding the impact on the efficiency of the tax system, the best policy recommendation would be to avoid tax rate cuts for all complements to CO_2 emitting goods and to look for strong substitutes instead.

Equation (5.12') further shows the importance of the tax bases for the total effect on the environment. Many proposals for green tax reforms suggest substituting the tax on the dirty good for labour taxes, i.e. to replace a broad-based tax by a narrow-based tax (Bovenberg and de Mooij, 1994: p. 1088) In this case, R_d / R_c becomes relatively small and the paradoxical case becomes less likely.

Hence, taxes on substitutes or on labour are good candidates for green tax reforms with respect to their ability to reduce carbon dioxide emissions. However, such tax reforms might have some undesired distributional consequences. If, in the absence of lump-sum transfers, the government wishes to compensate those members of the present generation who bear the carbon tax, the only way to ensure at least indirect compensation is to 'subsidise' complementary goods. Additional tax revenues from a carbon tax may therefore be used to reduce some narrow-based taxes and the paradoxical case cannot be ruled out a priori.

5.6 Conclusion

When discussing measures to fight global warming, care should be taken to properly evaluate the efficiency of green tax reforms. Within a distorted tax system, the introduction of carbon taxes does not necessarily reduce CO_2 emissions by the amount we would expect from focusing on the incentive effects of the carbon tax only. When looking at green tax reform proposals we have to be aware of the environmental impact of both the carbon tax and the accompanying measures taken by the government to guarantee revenue neutrality when looking at green tax reform proposals.

References

Ahmad, E. and N. Stern (1984) The Theory of Tax Reform and Indian Indirect Taxes. *Journal of Public Economics, 25, pp. 259-298.*

Baumol, W.J. and W.E. Oates (1988) *The Theory of Environmental Policy.* 2nd edition, Cambridge University Press, Cambridge, Mass.

Bovenberg, A.L. and R.A. de Mooij (1994) Environmental Levies and Distortionary Taxation. *American Economic Review, 84, pp. 1085-1089.*

Cline, W.R. (1992) *The Economics of Global Warming.* Institute for International Economics, Washington D.C.

F.D.P. (1993) Zukunftsicherung für den Wirtschaftsstandort Deutschland: Forderungen der F.D.P.. *Die Liberale 7-8, pp. 9-12.*

Goulder, L.H. (1995) Environmental Taxation and the Double Dividend: A Reader's Guide. *International Tax and Public Finance 2, pp. 157-184.*

Guesnerie, R. (1977) On the Direction of Tax Reform. *Journal of Public Economics 7, pp. 179-202.*

Houghton, J. T., G. J. Jenkins and J. J. Ephraums (Eds.) (1990) *Climate Change. The IPCC Scientific Assessment. Report Prepared for IPCC by Working Group 1.* Cambridge University Press, Cambridge.

Houghton, J. T., B. A. Callander and S. K. Varney (Eds.) (1992): *Climate Change 1992. The Supplementary Report to the IPCC Scientific Assessment.* Cambridge University Press, Cambridge.

Mayshar, J. (1990) On Measures of Excess Burden and Their Application. *Journal of Public Economics 43, pp. 263-289.*

Ng, Yew-Kwang (1980) Optimal Corrective Taxes or Subsidies when Revenue Raising Imposes an Excess Burden. *American Economic Review 70, pp. 744-751.*

Pearce, D.W. (1991) The Role of Carbon Taxes in Adjusting to Global Warming. *Economic Journal 101, pp. 938-948.*

Repetto, R. and R.C. Dower (1992) *Green Fees: How a Tax Shift Can Work for the Environment and the Economy.* World Resource Institute, Washington D.C.

Schöb, R. (1994) *Evaluating tax reforms in the presence of externalities.* Paper presented on the 50th Congress of the IIPF.

Schöb, R. (1995) *Ökologische Steuersysteme. Umweltökonomie und optimale Besteuerung.* Campus, Frankfurt, New York.

Sprenger, R. U. *et al.* (1993) *Umweltwirkungen des deutschen Steuer- und Abgabensystems und Möglichkeiten sowie Grenzen seiner stärkeren ökologischen Ausrichtung.* ifo-Gutachten für die Bayerischen Staatsministerien für Landesentwicklung und Umweltfragen, für Wirtschaft und Verkehr sowie Finanzen, München.

6 Analytic Solutions of Simple Optimal Greenhouse Gas Emission Models *

Stephen C. Peck and Y. Steve Wan
Electric Power Research Institute
P.O. Box 10412, Palo Alto
CA 94303, USA

Abstract

We demonstrate analytically the characteristics of optimal emission reduction policies by utilizing simple greenhouse gas emission cost reduction models for various climate damage functions. We also show how to incorporate these emission models into a decision framework to evaluate the value of information for key parameters. Comparative numerical examples are given.

6.1 Introduction

There has been a great deal of interest in deriving the optimal path of greenhouse gas emissions; see Nordhaus (1982 and 1990), Peck and Teisberg (1992), Kosubod, South, Daly, and Quinn (1991). Frequently, the optimal solution of these models implies a shadow price of emissions that is rising over time and it has been suspected that the problem bears a close relationship to the Hotelling problem of the optimal depletion of an exhaustible resource. In section 6.2 we specify a simple model of intertemporal cost minimization where cost has two components, an emission reduction cost, which depends on the current rate of emission reduction and an environmental damage cost which depends on the accumulated stock of an atmospheric pollutant. Due to the simplicity of the specification we are able to show analytically that such a specification admits several solutions depending on the form of the environmental damage function. If that function is linear, the optimal shadow price is to increase at the growth rate of the economy; if the damage function is stepped, the optimal shadow price increases exponentially at a rate equal to the sum of the interest rate and the greenhouse gas stock depreciation rate as in the Hotelling model.

In the third section we use the model in section 6.2 to demonstrate how a simple decision framework can be utilized to evaluate the expected value of perfect information for key

* This paper does not represent the views of EPRI or of its members. The authors are solely responsible for any errors. Correspondence should be directed to the address below.

NATO ASI Series, Partnership Sub-Series, 2. Environment – Vol. 14
Economics of Atmospheric Pollution
Edited by Ekko C. van Ierland and Kazimierz Górka

parameters. The fourth section provides an example that compares carbon shadow prices under linear and stepped damage functions.

6.2 A Greenhouse Gas Emission Reduction Model

The greenhouse gas emission reduction problem can be modeled as an optimal control problem. One possible formulation is to minimize the discounted sum of control and damage costs.

$$\begin{aligned} &\text{Minimize} \quad \int_0^\infty \left[-\beta \ln(1 - \frac{x}{\mu_0 e^{gt}}) + h(D) \right] e^{-rt} dt \\ &\text{subject to} \quad \dot{D} = \mu_0 e^{gt} - x - \delta D, \quad \text{and} \quad x \geq 0. \end{aligned} \tag{6.1}$$

where x is the non-negative control variable representing the greenhouse gas emission reduction level,

β is a parameter that reflects the rate of increase in control cost as reduction increases,

μ_0 is the unconstrained greenhouse gas emission level at time 0,

g is the common growth rate of the unconstrained greenhouse gas emission of the economy,

D is the difference between the current greenhouse gas mass in the atmosphere and the preindustrial mass,

δ is the depreciation rate for the greenhouse gas mass difference,

r is the real interest rate, and

$h(D)$ is the damage function resulting from the greenhouse gas mass difference D.

The emission reduction cost function exhibits the property that

- if there is no control ($x = 0$), control cost is zero;
- control cost increases asymptotically to infinity when the reduction level approaches to $\mu_0 e^{gt}$; and
- as the unconstrained emission increases over time, it becomes less expensive to reduce the same amount of emissions (representing technical change).

We examine the behavior of the control variable x and the marginal cost of emission reduction (emission tax) with respect to two damage functions $h(D)$. We consider a linear damage function with an expanding economy in section 6.2.1. In section 6.2.2, we assume the damage function takes the shape of a step function and the model becomes the Hotelling exhaustible resources problem. The characteristics of the optimal control corresponding to the linear damage function are very different from those of the optimal control corresponding to the stepped damage function.

6.2.1 Linear Damage Function

The linear damage function with an expanding economy can be expressed as

$$h(D) = Ke^{gt}D,$$

where K is a constant and g is the growth rate of the economy. This reflects the consideration that as the economy grows, the potential damage increases with the size of the economy. To solve for the optimal control x, we first form the current value Hamiltonian H

$$H = -\beta \ln(1 - x/(\mu_0 e^{gt})) + Ke^{gt}D + \theta(\mu_0 e^{gt} - x - \delta D).$$

The optimal control satisfies

$$\frac{\partial H}{\partial x} = \frac{\beta}{(1 - x/(\mu_0 e^{gt}))\mu_0 e^{gt}} - \theta = 0. \tag{6.2}$$

The minimum principle implies that

$$\dot{\theta} - r\theta = -\frac{\partial H}{\partial D} = -(Ke^{gt} - \theta\delta).$$

The general solution of θ takes the form

$$\theta = \frac{K}{(r + \delta - g)} e^{gt} + Ae^{(r+\delta)t}. \tag{6.3}$$

The condition

$$\lim_{t \to \infty} e^{-rt}\theta = 0$$

implies that $A = 0$ and $r > g$. Substituting θ into expression (6.2), we have

$$x = \mu_0 e^{gt} - \frac{\beta(r + \delta - g)}{K} e^{-gt}. \tag{6.4}$$

And D can be expressed as

$$D = \begin{cases} \frac{\beta(r+\delta-g)}{K(\delta-g)} e^{-gt} + Be^{-\delta t}, & \text{if } \delta \neq g \\ \frac{\beta(r+\delta-g)}{K} te^{-gt} + Be^{-gt}, & \text{if } \delta = g \end{cases}$$

where B is determined by the boundary condition $D(0) = D_0$.

The solution indicates that emission reduction increases exponentially at the rate g and approaches the unconstrained emission level asymptotically. The rate of increase in control cost, β, has no impact on the shadow price θ. It will, however, affect the optimal emission reduction level. For the special case of $g = 0$, the shadow price is constant, the optimal emissions reduction is constant and the atmospheric mass difference approaches $\beta(r + \delta)/K\delta$.

An interesting issue from the policy implementation point of view arises from the observation that β appears in (6.4) but not in (6.3). If we were to adopt a policy of curtailing emissions directly, we need to know β. However, if the policy were the imposition of the emission tax expressed in (6.3), the knowledge of β becomes irrelevant. Consequently the value of information, to be discussed in more details in section 3, for the parameter β would be very different based on the policy form to be implemented.

6.2.2 Hotelling Results

The stepped damage function is depicted graphically in Figure 6.1. The unconstrained emission is assumed to remain constant at μ_0.

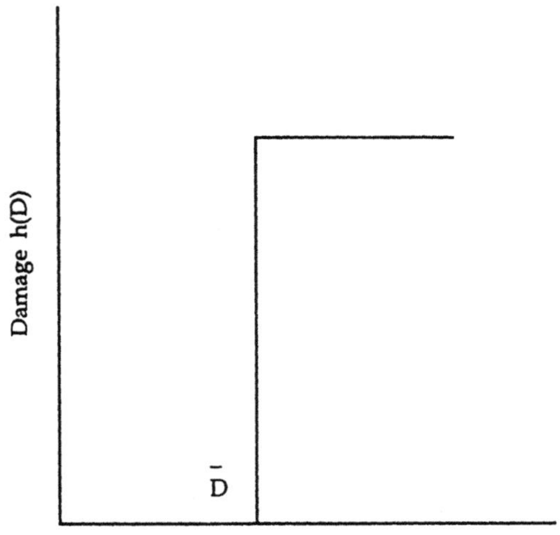

Figure 6.1 Damage function - step function.

The optimal control stragegy is to keep the incremental greenhouse gas mass at or below the threshold value, $\bar{D}$, of the step function. The optimal control problem takes the form

$$\begin{aligned} \text{Minimize} \quad & \int_0^T -\beta \ln(1 - \frac{x}{\mu_0}) e^{-rt} dt + \int_T^\infty -\beta \ln(1 - \frac{x_T}{\mu_0}) e^{-rt} dt \\ \text{subject to} \quad & \dot{D} = \mu_0 - x - \delta D \\ & D(0) = D_0 \\ & D(T) = \bar{D}. \end{aligned} \tag{6.5}$$

Choosing T that minimizes the objective function in (6.5) implies that

$$x_T = \mu_0 - \delta \bar{D}.$$

And the objective function in (6.5) can be written as

$$\text{Minimize} \quad \int_0^T -\beta \ln(1 - \frac{x}{\mu_0}) e^{-rt} dt - \beta \ln(\frac{\delta \bar{D}}{\mu_0}) e^{-rT}/r.$$

The above formulation is a variant of the well-known Hotelling model for exhaustible resources. The optimality conditions are

$$\frac{\partial H}{\partial x} = \frac{\beta}{(1 - x/(\mu_0))\mu_0} - \theta = 0. \tag{6.6}$$

and

$$\dot{\theta} - r\theta = -\frac{\partial H}{\partial D} = \theta\delta. \tag{6.7}$$

From (6.7) we have

$$\theta = Ae^{(r+\delta)t}$$

Plugging x into (6.6), we have

$$x = \mu_0 - \frac{\beta}{A} e^{-(r+\delta)t}$$

where A is determined by the boundary conditions in (6.5).

The Hotelling results suggest that the shadow price is characterized by positive exponential growth until x_T is reached when it becomes constant. In the case of linear damage function, the shadow price, expressed in (6.4), is constant if the growth rate g is 0.

6.3 Expected value of perfect information EVPI

The models in previous sections assume that all parameters are deterministic. These parameters, however, are subject to many uncertainties. One frequently asked question is: what is the benefit to have some uncertainties resolved in a timely manner. In addition, what are the key factors that would make the early resolution attractive; see Peck and Teisberg (1993) and Kolstad (1996). We use the model in section 6.2 with linear damage function to demonstrate how to get the expected value of perfect information (EVPI) for key parameters. The decision tree is represented in Figure 6.2. The parameter T denotes the time that the uncertainty is resolved. Without loss of generality, we derive the EVPI by focusing on the expected value of obtaining information at the beginning of the planning problem as opposed to never.

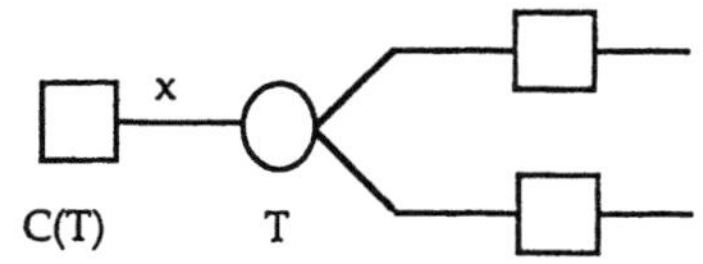

Figure 6.2 Decision tree with respect to timing.

Rewriting the model (6.1) with the linear damage function, we have

$$\begin{aligned} &\text{Minimize} \quad \int_0^\infty \left[-\beta \ln(1-\frac{x}{\mu_0 e^{gt}}) + K e^{gt} D\right] e^{-rt} dt \\ &\text{subject to} \quad \dot{D} = \mu_0 e^{gt} - x - \delta D, \quad \text{and} \quad x \geq 0. \end{aligned} \tag{6.8}$$

Assume that the uncertainty of interest is the unit damage cost K. Let P_h and P_l be the probabilities that the unit damage costs are K_h and K_l, respectively. If the state of the world (K_h or K_l) information were to become available at time T, then the objective function of (6.8) can be written as

$$C(T) = \int_0^T \left[-\beta \ln(1-\frac{x}{\mu_0 e^{gt}}) + \bar{K} e^{gt} D\right] e^{-rt} dt + \sum_{i=h,l} P_i \int_T^\infty \left[-\beta \ln(1-\frac{x_i}{\mu_0 e^{gt}}) + K_i e^{gt} D_i\right] e^{-rt} dt$$

where $\bar{K}=P_h K_h + P_l K_l$. The difference of $C(T_1)$ and $C(T_2)$ is the EVPI of having information available at time T_1 in contrast to information available at time T_2. We shall examine EVPI(0,∞) $= C(\infty) - C(0)$ and see how different components of the total cost (control and damage costs) contribute to EVPI(0,∞). We will also investigate how various parameters affect the EVPI(0,∞).

Recall that we are able to get explicit solution for the model if the damage function is linear, as demonstrated in section (6.2.1). We focus on the case that $\delta \neq g$. Hence $C(0)$ can be expressed as

$$P_h(\text{control}(K_h) + \text{climate-damage}(K_h)) + P_l(\text{control}(K_l) + \text{climate-damage}(K_l)) \tag{6.9}$$

where

$$\text{control}(K) = \frac{\beta}{r} \ln \frac{K\mu_0}{\beta(r+\delta-g)} + \beta \int_0^\infty 2gte^{-rt} dt$$

and

$$\text{climate-damage}(K) = \frac{\beta(r+\delta-g)}{r(\delta-g)} + \frac{K}{(r+\delta-g)}(D_0 - \frac{\beta(r+\delta-g)}{K(\delta-g)})$$

$C(\infty)$ can be written as

$$\text{control}(\bar{K}) + \text{climate-damage}(\bar{K}). \tag{6.10}$$

The difference between the control cost component of (6.9) and (6.10) is

$$\begin{aligned} &\frac{\beta}{r}(\ln \bar{K}\mu_0 - P_h \ln K_h\mu_0 - P_l \ln K_l\mu_0) \\ &= \frac{\beta}{r}(\ln(P_h K_h + P_l K_l) - \ln(K_h^{P_h} K_l^{P_l})) \\ &\geq 0 \end{aligned} \tag{6.11}$$

This implies that the earlier we have the information, the less we need to spend on the emission reduction. Their difference will contribute postively to EVPI(0,∞). The difference in climate damage can similarly be obtained and shown to have a constant value 0. Hence the climate damage is independent on the timing of information on the parameter K.

To summarize, the EVPI(0,∞) for the parameter K is positively correlated to

i) the difference between the unit damage costs $(K_h - K_l)$,

ii) the rate of control cost increase β,

and negatively correlated to the discount rate (r). And for given K_h and K_l, EVPI(0,∞) is at maximum when $P_h - P_l = \frac{2}{\ln K_h - \ln K_l} - \frac{K_h + K_l}{K_h - K_l}$. One can show that the right hand side of the above expression is always non-positive if $K_h > K_l$. Thus the value of information is usually higher if the true K is more likely to be K_l.

The value of information for the parameter β can be similarly obtained. Let π_h and π_l be the probabilities that the parameter β are β_h and β_l, respectively. Assuming that the optimal policy takes the form of emission reduction, then the EVPI for β can be expressed as

$$\frac{1}{r}(\pi_h \beta_h \ln \beta_h + \pi_l \beta_l \ln \beta_l - \bar{\beta} \ln \bar{\beta}) \geq 0$$

where $\bar{\beta} = \pi_h \beta_h + \pi_l \beta_l$. Note that the EVPI only consists of terms related to the control cost. On the other hand, if we were to adopt the tax policy, then the EVPI for β would be zero.

6.4 Numerical example

We provide a numerical example to compare the behaviors of the emission shadow prices under linear and stepped damage functions - all other conditions being equal. First we state the calibration assumptions for a future year that can be thought of as the middle of the next century.

a) emission rate: $\mu = 15.010^9$ tons/year, and initial atmospheric stock $D_0 = 45010^9$ tons.

b) control cost: At 1010^9 tons/year emission reduction, the cost would be 4% of the GWP (gross world production). GWP is assumed to be 5.3810^{13} \$/year. The parameter β is calibrated as

$$-\beta \ln(1 - (1010^9)/(1510^9)) = 0.045.3810^{13}$$

$$\Rightarrow \qquad \beta = 1.9610^{12}$$

c.1) damage cost - linear damage function: At 0.4510^{12} tons, the cost would be 1.5% of the GWP. We have

$$K(0.4510^{12}) = 0.0155.3810^{13}$$

$$\Rightarrow \qquad K = 1.79$$

c.2) damage cost - step damage function: We examine three cases with threshold values ($\bar{D}$) at 500, 600, and 700 billion tons respectively.

d) economy growth rate: g is 0.0% since we wish to compare the linear case with the Hotelling case (for which there is no growth).

e) discount rate: r is assumed to be 5.0%.

f) depreciation rate: δ is taken as 2.0%.

In the linear case, the emission shadow price equals

$$v_l = \frac{K}{(r + \delta - g)}.$$

In the Hotelling case, the emission shadow price equals

$$v_h = Ae^{(r+\delta)t},$$

where A is determined by the boundary conditions. Figure 6.3 depicts the emission shadow price trajectories of the two cases.

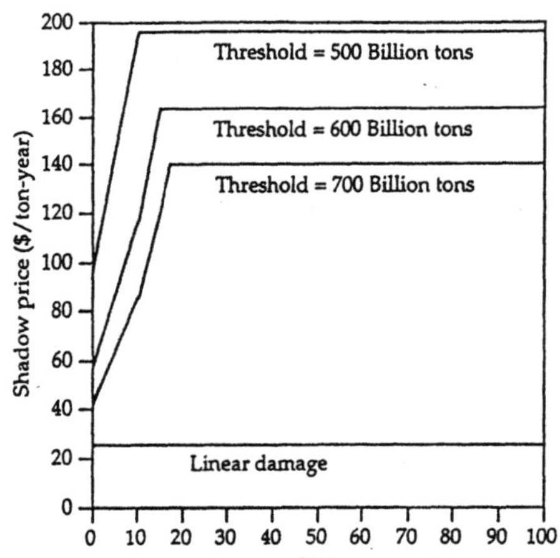

Figure 6.3 Shadow prices

6.5 Conclusion

We presented two simple greenhouse gas emission reduction models to demonstrate the behaviors of the optimal emission reduction policies for climate damage functions assuming different functional forms. For a linear damage function, the shadow price and the optimal emission reduction increase at the growth rate of the economy. In the special case of no growth in the economy, the shadow price is constant.For the step damage function as depicted in Figure 6.1, the optimal reduction policy is characterized by positive exponential growth at a rate equal to the sum of the interest rate and the greenhouse gas stock depreciation rate.

We demonstrated how to incorporate the emission reduction models into a decision framework to evaluate the EVPI in obtaining information for key parameters. We also showed the factors that contribute to the value of information by way of examples.

References

Nordhaus, William D. (1982) How Fast Should We Graze the Global Commons? *American Economic Review, May 1982.*

Nordhaus, William D. (1990) To Slow or Not To Slow: The Economics of the Greenhouse Effect? *International Energy Workshop, MIT.*

Peck, Stephen C. and Teisberg, Thomas J. (1992) CETA: A Model for Carbon Emission Trajectory Assessment, *Energy Journal, 1992.*

Peck, Stephen C. and Teisberg, Thomas J. (1993) Global warming uncertainties and the value of information: An analysis using CETA, *Resource and Energy Economics, 1993.*

Kosubod, R.F., South D.W., Daly T.A., and Quinn K.G. (1991) Tradeable CO2 Emission Permits for Cost-effective Control of Global Warming, Presented at *13th Annual North American Conference, International Association for Energy Economics, November 1991.*

Kolstad, C.D. (1996) Learning and Stock Effects in Environmental Regulation: The Case of Greenhouse Gas Emissions, forthcoming, *Journal of Environmental Economics and Management.*

7 The Design of Cost Effective Ambient Charges under Incomplete Information and Risk

Youri Ermoliev, Ger Klaassen and Andries Nentjes*
International Institute for Applied Systems Analysis
A-2361 Laxenburg
Austria

Abstract

This contribution discusses a dynamic approach for setting emission charges such that ambient standards at receptors are respected and costs are minimized even if the environmental agency has no information about the costs of emission control. The proof, which is based on non monotonic optimization procedure demonstrates that starting from an arbitrarily chosen set of ambient charges the adjustment process converges to the cost effective vectors of charges and induced emissions. After presenting a basic deterministic case, the adjustment process is extended to a model where costs and transport coefficients are stochastic. Next to that penalties for sources that emit more than they have reported are integrated in the model.

7.1 Introduction

Past experience and economic analysis both point out that environmental pollution creates a social problem that cannot be solved by the market. The public good or, more appropriate, public bad property of large scale pollution makes it impossible to organize complete environmental markets with private demand for and private supply of pollution control. Normally national or regional government fill the gap by formulating their (public) demand for pollution control and confronting polluters with these demands.

* University of Groningen, Groningen, The Netherlands.

NATO ASI Series, Partnership Sub-Series, 2. Environment – Vol. 14
Economics of Atmospheric Pollution
Edited by Ekko C. van Ierland and Kazimierz Górka

Ideally the acceptable (or optimal) level of pollution is determined by balancing the marginal costs and benefits of reducing pollution loads. Unfortunately the calculation of environmental damage and therefore the benefits of cleaning up is fraught with difficulties and often impossible, since reliable dose-effect relations cannot be established, or monetary evaluation of physical environmental effects makes little sense. In this paper we shall neglect the problem of finding the safe level of public demand for environmental quality and concentrate on the question of how to translate a given public demand for environmental quality (for example in the form ambient air quality standards within a spatial system) in demands on polluters, which can be viewed as suppliers of pollution control and therefore ultimately of environmental quality. The criteria used to assess the solution to the problem are environmental effectiveness (aspired levels of environmental safety should be realized) and cost effectiveness: the cost of supplying the environmental quality that is in demand should not be higher than strictly necessary.

The conventional method to control pollution is to collect necessary data and give strict orders to sources how much emission control and residual emissions to 'produce'. In other words, to impose emission standards.

When the environmental agency is fully informed about potential emissions, emission control cost functions and transport of emissions to receptors (the transport coefficients) is deterministic, the problem of finding the vector of emission standards that meets given ambient (air quality) standards in a cost effective way can be solved straightforward. The essence of the approach is to formulate the pollution control model as minimization of a total emission control cost function subject to (ambient) standards imposed as receptor points; see for example IIASA's RAINS model (Alcamo *et al.*, 1990). The model calculates emission standards for sources that meet ambient standards at receptors at minimum cost. The approach assumes a fully informed 'central planner' and a regulatory authority that succeeds completely in maintaining the imposed emission standards. When these conditions are not fulfilled in the real world it is not likely that the actual outcome of the policy is cost-efficient.

An alternative policy to translate public demand for environmental quality in a cost effective supply of emission control by polluters is to simulate a kind of market. Polluters can acquire permission to pollute (a kind of property right) by paying a price for each unit of released pollutant. By using classical duality results, Tietenberg (1978) has demonstrated that there exists a vector of ambient charges at receptors which would induce emissions of sources to meet ambient standards at minimum cost. The simple structure of the pollution transport equation allows to 'translate' the cost effective ambient charges into cost effective emission charges.

Yet there is a bottleneck which has prevented the actual application of cost effective emission charges to control ambient concentrations at multiple receptors until now. Designing cost effective emission charges is equivalent to the solution of the dual problem asking for the same full information that is needed for the calculation of cost effective emission standards (the primal problem), including information on emission control cost functions. This begs the question what the advantages are of using price signals in the emission charges compared with emission standards.

An alternative approach for setting emission charges is a procedure in which the authority starts with an arbitrary set of emission charges, monitors the discrepancy between rate in successive steps in such a way that excess pollution disappears. If such a system were practicable and cost effective the authority would not be in need to have knowledge of the emission control cost function and thus the information content needed to apply a system of cost effective emission charges would evidently be lower than for the implementation of cost effective emission standards.

The difficulty is that it is by no means clear that the solution to the dual problem can be found in practice simply by such a trial and error mechanism. Baumol and Oates (1971) have considered the possibility of such a mechanism for the cost-effective control of pollution, but only for the case of one receptor. It seems to be the generally accepted view today that with multiple receptors such an iterative adjustment procedure would not guarantee that the resulting allocation of emission control is cost effective (Bohm and Russell, 1985).

However, in an earlier paper (Ermoliev *et al.*, 1993) we have argued that a charge adjustment procedure can be designed that achieves ambient standards at multiple receptors at minimum cost and that in our view is suitable for practical use. We define the gap between actual concentration and the target concentration (the ambient standards) at receptors as 'excess pollution function'. We assume that the agency is able to monitor the values of this function and that there is agreement between the agency and polluters on the pollution transport model. According to the proposed procedure the agency separates the design of cost effective charges in two stages: first a learning phase in which no control measures are taken and a second stage of actual implementation.

In the first stage the agency organizes a kind of auction among polluters in the following way. An initial vector of emission charges is announced. The sources then state their emission levels. It is assumed that the agency is able to verify whether actual emissions (in stage two) are in accordance with announced emission. Under this assumption one can expect that sources will try to find the emission level that given the emission charge proclaimed by the authority (the auctioneer) minimizes their total cost (of emission control plus expenditure on charges). From the emission levels proposed by sources the agency in its turn derives the excess concentrations at receptor points. It revises its prices and announces a new set of charges in a second round, and so on.

We assume that in this whole procedure sources choose emission levels such that the marginal cost of emission control equals the emission charge. The agency adjusts current ambient charges in each round proportionally to excess pollution concentrations.

Efficient communication would be made possible by organizing the auction on a network of computers that connects sources and the agencies. The 'dialogue' between computers is repeated until the vector of cost effective charges has been discovered. In the stage of implementation the ambient charges derived in the first stage are imposed as prices that actually have to be paid. The levels of (cost effective) emissions proposed by sources in response to these charges are also imposed and enforced as emission quota per source, by way of appropriate verification procedures and penalties in case of violation emission commitments.

The upshot of the whole argument is that the organization of a kind of market where sources can supply their emission control (and can demand residual emissions) is feasible as a substitute for command and control systems.

The aim of this paper is manyfold. First of all we give the formal proof that the proposed adjustment process converges to an equilibrium solution and that in this way the environmental agency is able to determine a cost effective set of ambient charges without knowledge of pollution control costs. The proof is based on the use of non differentiable optimization tools. Since this field is discussed only in special publications, the proof is important for the understanding of the adjustment procedure. It will serve as a base for the extensions we shall give. In contrast with the basic case (the deterministic model) where the assumption is that the environmental agency has certain knowledge of diffusion of emission (the matrix of transfer-coefficients) the extensions introduce cases where transfer coefficients are random. Whereas the deterministic models ensure that concentrations of pollutants at receptors are always below the ambient standards, this is no longer the case in stochastic models. There we have to take into account the possibility that environmental limits will be surpassed on taking into account the risks that are involved. In the context of such stochastic models again the question has to be answered what type of instrument should be used to realize environmentally and cost effective emissions.

This contribution is organized in 7 sections. In Section 7.2 the basic deterministic model of the pollution control problem is discussed. In this model transfer coefficients and ambient concentration are known and of a deterministic nature. Section 7.3 contains a general discussion of the adjustment mechanism and its convergence in the deterministic model. In section 7.4 we proceed with formulating stochastic models that allow discussion of economic mechanisms for stochastic pollution control in the presence of environmental risks. We specify the feasibility concept of an emission strategy since constraints on deposition levels can be violated in the case of unfavorable conditions in principle for any realistic emission policy. The design of a cost effective emission standard policy ensuring that the concentrations at receptors remain within standards with a given frequency per unit of time (chance constrained emission standard policy) was studied by Lohani and Thanh (1978), Ellis *et al.* (1985) and (1986), Guldmann (1986) and Fuessle *et al.* (1927). In these studies it is assumed that the regulating emission agency has full information on emission control costs and on the distributions of random variables. The stochastic control model proposed in Section 4 allows to account for non-linear interactions between emission levels, random pollution and risks of surpassing critical threshold of environmental absorption capacity. Still it allows to design the cost effective emission charges policy in a manner similar to Section 7.3. The mechanism for adjusting emission charges is discussed in Section 7.5. The procedure uses random excess pollution concentrations generated by actual emission and random characteristics of pollution dispersion. The main mechanism for discovering the cost effective (equilibrium) vector of ambient and derived emission charges is again a computerized dialogue between the agency and sources. In addition to Section 7.3 we introduce reported and actual emission levels, which provides a possibility to also discuss verification mechanisms. Section 7.6 illustrates how the proposed adjustment mechanisms function and Section 7.7 gives the conclusions. Proofs are discussed in the appendix.

7.2 Deterministic pollution control

Instruments of pollution control policy have to be designed in such a way that they are environmentally and cost-effective. When direct regulation is applied the problem of finding the vector of emission limits per source that meets the criteria can be captured in the following model.

Let $i = 1,\dots,n$ be sources of emission and $j = 1,\dots,m$ receptors. Each polluter i emits a single pollutant x_i. The transport of emissions $x = (x_1,\dots,x_n)$ to receptors is described by a transfer matrix $H = \{h_{ij}\}, i=1,\dots,n, j=1,\dots,m$, where h_{ij} stands for the contribution made by one unit of source i emission to the concentration of pollutants at receptor j. Ambient concentrations q_j at the receptor points $Q = (q_1,\dots,q_m)$ impose constraints on emission rates such that the total deposition at j does not exceed q_j:

$$\sum_{i=1}^{n} x_i h_{ij} \leq q_j, j=1,\dots,m \tag{7.1}$$

$$x_i \geq 0, i=1,\dots,n \tag{7.2}$$

The environmental authority's problem is to achieve target concentrations Q at minimum total cost for polluters: that is to choose the vector $x = (x_1,\dots,x_n)$ to minimize the total cost

$$\sum_{i=1}^{n} f_i(x_i) \tag{7.3}$$

subject to constraints (7.1)-(7.2).

The solution of this problem $x^* = (x^*_1,\dots,x^*_n)$ defines the cost effective set of emission levels which can be imposed as emission standards in the form of emission limits (or non-tradeable emission quota) per source. The implementation of such a policy of direct regulation requires that the regulatory authority has full information on parameters h_{ij}, q_j and all cost functions $f_i(x)$.

Instead of setting emission limits per source the environmental agency could consider the use of emission charges. Let's assume $f_i(x)$ is a convex function. By using classical duality results Tietenberg (1978) has demonstrated that there exists a vector $\lambda = (\lambda_1,\dots,\lambda_m)$ of shadow prices which can be considered as ambient charges at receptors which corresponds with the vector of cost-effective emission standards. The simple linear structure of the pollution transport equations allows the calculation of a vector of emission charges u_i. The emission charge can be written as a function of transfer coefficients and the shadow prices λ_j, $j=1,\ldots,m$ at the receptors:

$$u_i = h_{i1}\lambda_1 + h_{i2}\lambda_2 + \ldots + h_{im}\lambda_m \tag{7.4}$$

If optimal emission charges $u_1 = u_1^*$ associated with optimal $\lambda_j = \lambda_j^*$ are imposed they will induce sources to adjust emissions in a way which minimizes total expenditure on pollution control and charges:

$$f_i(x_i) + u_i x_i = min,\ x_i \geq 0 \tag{7.5}$$

The above results are well known. But they don't answer the main question, that is, how the environmental authority can determine the cost-effective charges $u_i = u_i^*$ without knowledge of the cost functions when there is more than one receptor (Bohm and Russel, 1985). In principle the authority could easily find a vector of high enough emission taxes to meet the target concentrations, but there is, however, no guarantee that the cost of emission control is minimized.

7.3 An artificial market mechanism

In this section we describe an adjustment mechanism that is capable to find the cost-effective vector of emissions through the adaptation of the emission charges in successive steps. The proposed procedure decomposes the pollution control problem into two types of decision problems. The first choice problem is that of the environmental authority having to decide on the ambient charge rates (and 'translate' them into emission charges) and adjust the rates, given information on the discrepancy between actual and target concentrations in receptors. The procedure does not require information on the total cost of controlling emissions. The proof will be given that actually the environmental authority can concentrate only on monitoring the excess concentration, that is the gap between actual and target concentrations

$$\Gamma_j(x) = \sum_{i=1}^{n} x_i h_{ij} - q_j\ ,\ j = 1,\ldots,m$$

without bothering about costs (see appendix).

The second optimization problem concerns polluters which choose their optimal emission level given the emission charge that is imposed by the environmental agency. The formal description of the adjustment mechanism is the following.

Suppose $\lambda^0 = (\lambda_1^0,\ldots,\lambda_m^0)$ is a vector of initial ambient charges on concentrations at receptor points and let $\lambda^k = (\lambda_1^k,\ldots,_m^k)$ be the vector of ambient charges at step k of the adaptive process. Each source adjusts its emission level x_i^k, $i=1,\ldots,n$ by minimizing its individual cost function $f_i(x) + u_i^k x$ at current values of ambient charges $\lambda_j^k, j = 1,\ldots,m$, which are translated into emission charges u_i^k by equation (7.4). Thus

$$u_i^k = \sum_{j=1}^m \lambda_j^k h_{ij}, i = 1,\ldots,n$$

The agency observes the ambient concentrations generated by the vector x^k; it calculates the excess concentration between actual and target concentrations $\Gamma_j(x^k)$ and adjusts the ambient charges in the next step according to the equation

$$\lambda_j^{k+1} = max\left\{0, \lambda_j^k + \rho_k\left(\sum_{im}^n x_i^k h_{ij} - q_j\right)\right\} \tag{7.6}$$

where $j = 1,\ldots,m$ and $k = 0,1,\ldots$. Therefore $\lambda_j^{k+1} = 0$, if $\lambda_j^k + \rho_k\left(\sum_{j=1}^n x_i^k h_{ij} - q_j\right) < 0$. The step size multiplier ρ_k is chosen as to ensure the convergence of the sequence λ^k, $k = 0,1,\ldots$ to the optimal dual variables $\lambda^* = (\lambda_1^*,\ldots,\lambda_m^*)$.

The appendix provides the proof that convergence takes place under a rather broad range of ρ_k; for example $\rho_k = c/k$, where c is an arbitrarily chosen positive constant, particularly $\rho_k = 1/k$. The proof is based on the observation that the direction of the movement $\Gamma(x^k) = \left(\Gamma_1(k^k),\ldots,\Gamma_m(x^k)\right)$ in (7.6) in a sense is a gradient of the following, in general, non-differentiable function (see appendix).

$$\gamma(\lambda) = min_{x\geq 0}\left[\sum_{i=1}^n f_i x_i + \sum_{j=1}^m \lambda_j \sum_{i=1}^m (x_i h_{ij} - q_j)\right] = L(x(\lambda), \lambda) \tag{7.7}$$

Procedure (7.6) can be interpreted as a kind of market system. Cost minimizing polluters independently adjust emissions to the current taxes u_i^k. The agency observes $\Gamma(x^k)$ and reacts as Walrasian auctioneer would. In case of excess concentrations at receptor j ($\Gamma_j > 0$) he will raise the price λ_j for users. In the other case the price will be lowered. Prices are signals for users of environmental space to adjust their emissions accordingly.

The procedure (7.6) can be modified easily to include any additional a priori information on possible ambient charges. For example, if we know an upper bound $\bar{\lambda}_j$, j = 1,...,m on λ_j^*.

It should be noted that in adjustment mechanism (7.6) value ρ_k,k = 0,1,... can not be chosen in order to decrease the total cost $\sum_{i=1}^n f_i(x_i)$ by passing from x^k to x^{k+1} since this function is unknown to the agency. The procedure does not lead to monotonic improvements of total costs since the choice of the step size multipliers $\rho_0,\rho_1,\ldots$ is not based on excess concentrations and not on the calculation of the total cost. Nevertheless the process enables to find optimal ambient charges. We also point out that the convergence takes place even for non-convex functions f(x),i=1,...,n.

7.4 Stochastic models

Sections 7.2 and 7.3 discussed the problem of cost effective pollution control in a deterministic 'environment'. Coefficients and parameters that determine emission control cost functions and dispersion of emission are fixed. In this section we drop this assumption and take into account that dispersion of pollution can be of a stochastic nature. Next to that we introduce the possibility that reported emissions diverge from actual emissions and that sources are penalized if an excess of actual emissions over reported emissions is detected. The change in underlying assumptions, bringing them more in agreement with the phenomena of the real world, affects the formulation of our cost minimization problem and its solution.

As to the costs of pollution control per source they depend on a number of factors next to the planned level of emissions and they can vary, thus affecting costs. Some are of a rather general kind like prices of inputs; others are more firm specific; for example composition and quantity of output (which affect potential emissions), types of fuel and other raw materials used and temporary breakdown of installed pollution control equipment. We shall assume that all these possible variations make for a stochastic cost function and that the source knows the probability distribution function. The polluter aims at minimizing an expected value cost function instead of a deterministic cost function.

The dispersion of pollutants by air is affected by factors as wind direction, wind speed and weather conditions, which are all variable. The concentration of effluents in water depends on the volume and velocity of the flow. Therefore coefficients h_{ij} of the deterministic model in section 7.3 are usually computed with Gaussian type diffusion equations. These models are run over all possible meteorological conditions, and the 'outputs' are then weighted by the frequencies of the meteorological inputs over a given time interval, yielding average transfer coefficients.

One option is to simplify and use average transport coefficients instead of stochastic transport coefficients, as was done in sections 7.2 and 7.3. It would ensure that on average pollutions concentrations in receptors do not exceed ambient standards. This is an acceptable approximation of reality when the effects of pollution vary in a smooth and continuous way with the size of pollution. 'Overshooting' of targets is then compensating by 'undershooting'. Unfortunately, for most environmental problems effects of pollution are not gradual. For example, acidification, radioactivity, ozone depletion are examples where pollution is relatively harmless until it reaches a sufficiently high level. When ambient standards have been set at or close to these critical levels or thresholds incidental excess pollution may cause a discontinuous deterioration in environmental quality, which even might be irreversible. Apart from this a further complication is that the ambient standards, set by the authority, itself may be uncertain in the sense that the critical level itself can vary, depending on weather conditions like sunshine and wind speed for processes of ozone formation. The fundamental question then is how to incorporate such risks in our approach. A natural improvement of deterministic models is the inclusion of probabilistic constraints on ambient standards that account for the random nature of pollution processes.

When an emission control policy has to be made in the presence of random transfer coefficients h_{ij} and actual ambient standards q_j we have to specify the feasibility concept of an emission strategy $x = (x_1,\ldots,x_n)$, since for any x some of the constraints

$$\sum_{i=1}^{n} h_{ij}x_i \leq q_j, \; j=1,\ldots,m \tag{7.8}$$

can be violated for concrete random realizations of h_{ij}, q_j, $i=1,\ldots,h$; $j=1,\ldots,m$. We assume that the environmental authority's problem is again to achieve (in a sense, which has to be specified in this section) target concentrations $Q = (q_1,\ldots,q_m)$ at minimum total expected cost for polluters.

$$\sum_{i=1}^{h} f_i(x_i) \tag{7.9}$$

First of all let us note that if we can account for the loss whenever the constraints are violated, then we can add the loss functions to the objective function. Thus assume that $g_j\left(\sum_{i=1}^{h} x_i h_{ij} - q_j\right)$ is the loss function associated with the excess concentration

$$\Gamma_j(x_1 w) = \sum_{i=1}^{h} x_i h_{ij} - q_j,\ j=1,\ldots,m$$

where w denotes the collection of all random variables $w = \{h_{ij}, q_j,\ i=1,\ldots,n,\ j = 1,\ldots,m\}$ that affect levels of excess pollution at receptors. Then the environmental authority's problem is to achieve emission levels $x = (x_1,\ldots,x_k)$ at minimum total social cost

$$F(x) = \sum_{i=1}^{n} f_i(x_i) + \sum_{j=1}^{m} d_j E g_j(\Gamma_j(x_i w)) \tag{7.10}$$

where E is the symbol of the expectations. The expectations $Eg_j(\Gamma_j(x_i,w)$ measure the violation of the environmental standards and environmental risks associated with that. The weighting coefficients $d_j \geq 0$ can be used to convert units of losses g_j into the unit of costs f_i. If impacts of pollution on environmental quality at receptors are continuous, the weights could be given a lower value. We don't take the view that quantification of the weights can be based on exact scientific calculations, but believe that the weights could be used to reflect the environmental authority's view of the acceptability of surpassing the actual environmental standards. (A more sophisticated method to take into account the asymmetry of discrepancies between actual concentrations and target levels (ambient standards) will be discussed hereafter and is captured in equations (7.14) to (7.17)). Thus the choice of the cost-effective emission levels in the presence of environmental risks is reduced to the minimization of function (7.10), reflecting the total cost of emission reduction plus costs associated with the violation of ambient standards.

If the environmental agency has information on costs $f_i(x_i)$ then the function F(x) can be minimized by using stochastic optimization tools, providing emission levels that can be imposed on sources as cost effective emission standards. However, from the discussion in section 7.2 follows that it is unrealistic to assume that functions $f_i(x_i)$ are known to the agency. The straight forward minimization of F(x) via the decomposition in independent subproblems for each source i is impossible since the risk function (second term of F(x)) is not separable with respect to variables $x_1,\ldots,x_n$, as it was the case with linear feasibility constraints (7.8) (see Section 7.3). The simplest strategy from the point of view of requirements on data is to impose so called "best available control technology" on sources. Obviously this policy leads to pollution reduction as much as is technically possible at any point in space and time. Unfortunately such a universal policy is expensive if it is applied generally, given the differences in weather characteristics, sensitivity in ecosystems, firm specific conditions, etc.

Minimization of function (7.10) leads to a socially cost-effective non-uniform emission standard policy. The main question is how the agency can organize the participation of sources in the minimization of this function. We shall address that problem in the next section (7.5). In this section we continue by first discussing how the environmental authority can respond to discrepancies between reported and actual emission (either on purpose or by accident) and a more thorough discussion is given of approaches to take into account the possibility that ambient standards are offended.

Any pollution control policy has to be enforced by sanctions, otherwise there will exist incentives for a source i to report to the agency a level of (reported) emissions y_i deviating from the actual level x_i. We assume that source i knows the penalty function Ψ_i, that is the fee the source has to pay for a positive difference $(x_i-y_i)_+ = \max\{0, x_i-y_i\}$. The function may also depend on some variables v which reflect impacts on actual and reported emissions that are not fully under control of the polluter. Therefore let $\Psi_i(x_i-y_i,\nu_i)$ be the penalty function of source i. Of course, it may be identical for all sources. We also assume that sources form their expectations $\varphi_i(x_i,y_i)$ concerning υ_i in order to choose their reported levels y_i of emissions. In general these expectations are different for different sources. Adding this element modifies the social cost function (7.10) to the following

$$F(x) = \sum_{i=1}^{n} C_i(x_i,y_i) + \sum_{j=1}^{m} d_j E g_j(\Gamma_j(x,w)) \qquad (7.11)$$

where

$$C_i(x_i,y_i) = f_i(x_i) + \varphi_i(x_i,y_i) \qquad (7.12)$$

So far we discussed the case where the agency is able to monitor and account for the losses and can evaluate them as costs, whenever the constraints (7.8) are violated.

Important approaches for a chance-constrained emission standard strategy have been analyzed by Ellis *et al.* (1986) and Guldmann (1986), Fuessle *et al.* (1927), Lohani and Thanh (1978). The strategy provides cost-effective emission levels minimizing total cost (7.9) while dealing with the following type of probabilistic constraints

$$G_j(x) = P_{\Gamma}\left\{\sum_{i=1}^{n} x_i h_{ij} \leq q_j\right\} \geq P_j,\ j=1,\dots,m \tag{7.13}$$

where levels P_j reflect risks of ambient standard achievement at receptor points. The probability P_j ensures that the concentration at receptor j remains below the ambient standard, with the frequency P_j per unit of time. For example, the constraint could require that ambient standard q_j is exceeded no more than once a year, or month, hour etc. The chance constraints destroy the linearity of the deterministic constraints (7.8) and introduce nonlinearity of random interactions between emission levels, meteorological conditions and ambient standards. Depending on the probability distribution of the variables $w = \{h_{ij}, q_j, i=1,\dots,n,\ j=1,\dots,w\}$, it may be necessary to reduce emissions for some sources considerably, contrary to the outcome under deterministic constraint (7.8) in mean. Of course, there is also the possibility to reduce emissions, at a very high cost, in order to insure that ambient standards are never violated $P_j=1,\ j=1,\dots,m$. The chance constraints (7.13) with $P_j \neq 1$ provide a more reasonable alternative, allowing for some violations while strictly controlling for the frequencies of the violations (risk indicators).

Unfortunately the chance constraints in general lead to multi-extremum and even discontinuous functions $G_i(x)$. For example, discontinuity of $G_i(x)$ occurs when random variables w are characterized by a finite number of "scenarios". In other words, when w has a discrete probability distribution, or when the values of w are generated via Monte-Carlo simulations. But instead of chance constraints, we may use constraints involving expectations of rather smooth functions producing convex constraints, and ensuring given levels of risks. Let us define similar to Klein Haneveld (1986) as $\Gamma_j(x_i w)_+ = \left(\sum_{i=1}^{n} x_i h_{ij} - q_j\right)_+$ the positive part of the excess concentrations $\Gamma_j(x_i w)$. The constraints can be defined as

$$\mathrm{E}(\Gamma_j(x_i w)_+) \leq r_j,\ j=1,\dots,m \tag{7.14}$$

where $r_j > 0$ are given mean values of permissible violations. Again r_j , $j=1,\ldots,m$ can be viewed as risk indicators associated with random violation of ambient standards.

When formulating probabilistic constraints (7.14), then instead of $\Gamma_j (x_i)_+$ we can take any convex function $U_j(\cdot,w), U_j(0,w) = 0$ to define risks of violation $u_j(\Gamma_j(x_i)_+,w)$ from $\Gamma_j(x_i) \leq 0$.

For example we can take $U_j(\Gamma_j(x_i)_+,w) = (\Gamma_j(x_i)_+)^2$ that ensures the resulting indicator to be a continuously differentiable function with respect to x for each w. In what follows we assume for the simplicity of notations that $U_j(\Gamma_j(x_i)_+,w) = (\Gamma_j(x_i)_+)^2$.

Now the cost-effective emission standards policy can be formulated as the minimization of total cost

$$\sum_{i=1}^{n} C_i(x_i,y_i) \tag{7.15}$$

while dealing with the following probabilistic constraints

$$G_j(x) = \mathrm{E}(\Gamma_j^2(x_i,w)_+) \leq r_j,\ j=1,\ldots,m \tag{7.16}$$

where $x_i \geq 0$, $y_1 \geq 0$ and r_j, $j=1,\ldots,m$ are given acceptable levels of risk indicators and $\Gamma_j^2(x_i)_+ = (\Gamma_j(x_i)_+)^2$.

If we assume that the agency possesses the information about cost functions $G(x_i,y_i)$ for all receptors, then the problem described by eqs. (7.15), (7.16) can be solved by using the stochastic optimization techniques (Ermoliev and Wets, 1988). The main complexity in this case is involved in the calculation of functions $G_j(x)$ and their gradients. The solution of problem (7.15)-(7.16) provides optimal levels y_i of reported emissions that have to be verified by the agency.

As was mentioned before the assumption that the agency has access to the information concerning cost function G is unrealistic. In the next section we examine a possible way to decentralize the problem (7.15)-(7.16) into independent sub-problems for sources and coordination of decentralized decisions by the agency in a way similar to the procedure presented in Section 7.3.

7.5 Adjustment mechanism

The problem described by eqs. (7.15)-(7.16) can be considered as a model ("laboratory") to test the feasibility of various pollution control policies in the presence of environmental risks reflected by probabilistic constraints.

The main question is how the environmental agency can minimize total cost (7.15) subject to constraint (7.16) without knowledge of the cost function $C_i, i=1,\ldots,n$. First of all, when all convexity and regularity conditions are valid, then it follows from the classical duality results for problem (7.15)–(7.16) that there exists a vector $\lambda^* = (\lambda_i^*,\ldots,\lambda_m^*)$ of shadow prices at receptors such that optimal levels of emissions x_i^*, y_i^*, $i=1,\ldots,n$ minimize

$$\sum_{i=1}^{n} C_i(x_i,y_i) + \sum_{j=1}^{m} \lambda_j^* (G_j(x) - r_j)$$

Functions $G_j(x)$ are continuously differentiable with gradients $\Delta G_j(x)$. Hence we can say that a necessary and sufficient condition for the optimality of (x_i^*,y_i^*) is that (x_i^*,y_i^*) minimizes the function

$$\sum_{i=1}^{n} C_i(x_i,y_i) + \sum_{j=1}^{m} \lambda_j^* \langle \Delta G_j(x*),x\rangle =$$
$$\sum_{i=1}^{n} [C_i(x_i,y_i) + \sum_{j=1}^{m} \lambda_j^* \frac{\partial G_j(x*)}{\partial x_i} x_i]$$

We can consider λ_j^* as ambient charges at receptors, and

$$U_i^* = \sum_{j=1}^{m} \lambda_j^* \frac{\partial G_j(x^*)}{\partial x_i},\ i=1,\ldots,n$$

as emission charges, which, in contrast to the deterministic case (eq. (7.4)), depend on the level of emissions x^* through the derivatives $\frac{\partial G_j}{\partial x_i}$ of the risk indicators. From the above follows that a vector of emission charges U_i^* exists. If charges U_i^* are imposed they will induce sources to adjust emissions in such a way that the sum of their control costs plus expenditure on emission charges is minimized:

$$C_i(x_i,y_i) + U_i^* x_i = min, x_i \geq 0, y_i \geq 0 \tag{7.17}$$

for i=1,...,n. It results in the cost effective emissions $x_1^*, y_i^*,\ i=1,...,m$ minimizing the total cost (7.15) subject to (7.16).

Let us describe now an adjustment mechanism that enables the agency to determine the vector of cost effective ambient charges λ_j^* and derived emission charges U_i^* without knowledge of the cost functions $C(x_1,y_1)$. As in the procedure described in section 7.3 the agency can restrict its task to monitoring the excess concentrations which now are random

$$\Gamma_j(x,w) = \sum_{i=1}^{n} x_i h_{ij} - q_j,\ j=1,...,m$$

The formal description of the adjustment mechanism runs as follows. Assume that the agency can calculate $G_1(x)$ and derivatives $\frac{\partial G_j(x)}{\partial x_i}$. For example the agency can estimate them from available observations of $\Gamma_j(x,w)$ as

$$G_j(x) \approx \frac{1}{N}\sum_{s=1}^{N} \Gamma_j^2(x,w^s)_+ \tag{7.18}$$

$$\frac{\partial G_j(x)}{\partial x_i} \approx \frac{1}{N}\sum_{s=1}^{N} \Delta\Gamma_j^2(x,w^s)_+ \tag{7.19}$$

$$\Delta\Gamma_j^2(x,w^s)_+ = \frac{1}{\Delta_s}\sum_{i=1}^{n} (\Gamma_j^2(x+\Delta_s e^i,w^s)_+ - \Gamma_j^2(x,w^s)_+)$$

where e^i is the unit vector of coordinate axe; $\Gamma_j(x,w^1),...,\Gamma_j(x,w^N)$ are independent observations of $\Gamma_j(x,w)$, j=1,...,m.

Note that these formulas provide estimates of $G_j(x)$ and $\frac{\partial G_j(x)}{\partial x_1}$ only by using observable excess concentrations $\Gamma_j(x,w)$. The proper procedure is using the general framework of stochastic optimization. For the sake of simplicity we restrict the discussion only to the case the environmental agency has exact knowledge of $G_j(x)_1\ \frac{\partial G_j(x)}{\partial x_i}$.

The agency may proceed then in a way similar to procedure (7.6). Suppose $\lambda^0 = (\lambda^0_1,\ldots,\lambda^0_m)$ is a vector of initial ambient charges on pollution concentration at receptors and let $\lambda^k = (\lambda^k_1,\ldots,\lambda^k_m)$ be the vector of ambient charges at step k of the adjustment process. The agency calculates the current emission charges (by using available information on excess concentrations as in (7.18), (7.19))

$$U_i^k = \sum_{j=1}^m \lambda_j^k \frac{\partial G_j(x^k)}{\partial x_i} \tag{7.20}$$

and submits this information to sources. Each source adjusts its reported emission level y_i^k to y_i^{k+1} by minimizing its total individual cost:

$$C_i(x_i,y_i) + U_i^k x_i,\ x_i \geq 0,\ y_i \geq 0 \tag{7.21}$$

Let $\overline{x_i^k}$ be the corresponding optimal value of x_i. The adjustment of actual emission level takes place with a precaution

$$x_1^{k+1} = x_i^k + \gamma_i^k\left(\overline{x_i^k} - x_i^k\right),\ i = 1,\ldots,n \tag{7.22}$$

where γ_i^k is a step size multiplier ($\gamma_i^k > 0$) used by i- to source at step k to adjust the current actual emission level x_i^k. The convergence of the adjustment mechanism takes place under a rather broad range of γ_i^k, for example $\gamma_i^k = c_i/k^\epsilon$ where c_i is an arbitrary chosen positive constant, and $0 < \epsilon < 1$.

Using information on excess concentrations $\Gamma_j(x^k,w)$ the agency calculates the expectations $G_j(x^k)$ and derivatives $\dfrac{\partial G_j(x^k)}{\partial x_i}$ and adjusts the ambient charges in the next step according to the rule that is similar to eq. (7.6):

$$\lambda_j^{k+1} = max\ \left\{0, \lambda_j^k + \rho_k(G_j(x^k) - r_j)\right\} \tag{7.23}$$

where $j=1,\ldots,m$ and $k=0,1,\ldots$ and ρ_k is a step size multiplier such that $\rho_k = c/k$ where c is an arbitrary chosen positive constant, particularly, $\rho_k = 1/k$.

In order to provide the intuition behind the procedure (7.20)-(7.23) we shall show that in the deterministic case this procedure is reduced to the adjustment mechanism proposed in section 3.

Assume that parameters h_{ij}, q_j in (7.8) are deterministic. Take in (7.17) $r_j=0$, $j=1,\ldots,m$. Then constraints (7.17) are reduced to the requirement $\Gamma_j^2(x,w)_+ = 0$ which is equivalent to

$$\sum_{i=1}^{n} x_i h_{ij} - q_j \leq 0$$

and (7.23) is reduced to procedure (7.6). Eq. (7.20) is also reduced to the eq. (7.4). In pollution control model (7.1)-(7.3) of section 7.3 the variables $x_i \equiv y_i$, $i=1,\ldots,n$. Therefore eqs. (7.20) can be dropped.

The convergence analysis of the procedure (7.20)-(7.23) is more complicated than the adjustment mechanism of section 7.3. The important feature of this procedure is the following. Although it is described in terms of functions $G_j(x), \frac{\partial}{\partial x_i} G_j(x)$ the actual information which is needed for the agency are observations of excess concentrations $\Gamma_j(x,w)$ only (see eqs. (7.18)-(7.19)). Estimates of exact values $G_j(x), \frac{\partial}{\partial x_i} G_j(x)$ obtained by using random values $\Gamma_j(x,w)$ are also random. Therefore the rigorous analysis of procedure (7.20)-(7.23) requires the general framework of stochastic optimization (Ermoliev and Wets, 1988). In the appendix we prove only the convergence of the version described in section 7.3 for the deterministic pollution control model and we outline the main ideas involved in the proof of procedure (7.20)-(7.23).

The interpretation of the adjustment mechanism as an artificial market system is similar to that in section 7.3. The agency calculates the current emission charges according to equation (7.20) where $\frac{\partial}{\partial x_i} G_j(x)$ in eq. (7.19) are calculated by using only values $\Gamma_j(x,w) = \sum_{i=1}^{n} x_i h_{ij} - q_j$. These values are monitored by the agency in on-line regime at the implementation stage, or they are generated by a Monte Carlo simulation of coefficients h_{ij} and q_j. The result $\Gamma_j(x,w)$ is known to the agency. The adjustment of ambient charges according to eq. (7.23) uses also only information about $\Gamma_j(x,w)$ as in (7.18). The adjustment mechanism (7.21)–(7.23) admits two levels of emissions: reported y^k and actual x^k. Different verification procedures can be incorporated in the definition of expected penalties $\Phi_i(x_i,y_i)$ for costs $C_i(x_i,y_i)$. We don't specify them and assume only that sources participate in the procedure and that they are interested to reach optimal emission levels x_i, y_i. It can be shown that if sources don't adjust the actual emission levels x_i in a way as is suggested by eq. (7.22), then the cost effective emission levels can not be achieved in general.

7.6 Numerical results

Acidification is one of the major problems in Europe and the Netherlands. Ammonia emissions are a major source of acid rain in the Netherlands. These emissions mainly result from livestock farming and fertilizer use and are generally transported over short distances (50% is deposited within 100 kilometres of the source). This implies that the major sources of ammonia deposition in the Netherlands are in the Netherlands, Belgium, France, Western Germany, Ireland, Luxembourg and the United Kingdom which contribute to four receptor areas (grid size 150 x 150 km). The Netherlands' policy is to reduce acid deposition to 2400 equivalents of acid/hectare in the year 2000. After subtracting the expected contribution from sulphur and nitrogen oxides in the year 200, targets for ammonia deposition can be formulated for each grid.

For the adaptive charge mechanism data are needed on transfer coefficients and deposition levels. These levels can be simulated by solving subproblems (7.5) for each source and given current taxes u_i or λ_j calculated according to eq. (7.4)–(7.6). Transfer coefficients for ammonia are based on European Program for Monitoring and Evaluation (Sandnes and Styve, 1992). The transfer coefficients for the four Dutch receptors are based on the average meteorology for 1985, 1987, 1988, 1989 and 1990, which clearly show the short travel distances of ammonia. The costs for controlling ammonia emissions are based on the RAINS model of IIASA (Alcamo *et al.*, 1990). RAINS stands for Regional Acidification Information and Simulation. RAINS distinguishes the following options for controlling ammonia emissions: low ammonia manure application, ammonia poor stable systems, covering manure storage, cleaning stable air, low nitrogen fodder and industrial stripping. For each country the potential and costs of their techniques are calculated accounting for country- and technology-specific factors (Klaassen, 1991). These options are then compiled in national cost functions which rank the options according to their marginal costs and volume of emissions removed.

Two simulations were carried out:

1. the agency starts with initial deposition charges of zero (scenario 1)
2. the agency starts with a deposition charge of 100,000 DM/kton NH_3 deposited in grids 3 and 4 (scenario 2).

The reason for scenario 2 is that the agency knows immediately that without any control the deposition targets at receptors 1 and 2 are already met. Figures 7.1 and 7.2 show the values of both the total (annual) pollution control costs and the Lagrange function as a function of the number of iterations. Figure 7.1 shows that total cost of control plus emission charges (the Lagrange function) converges fast to optimal value. Obviously, starting from a zero charge would initially lead (see step 2) to very high charges. This is especially so for

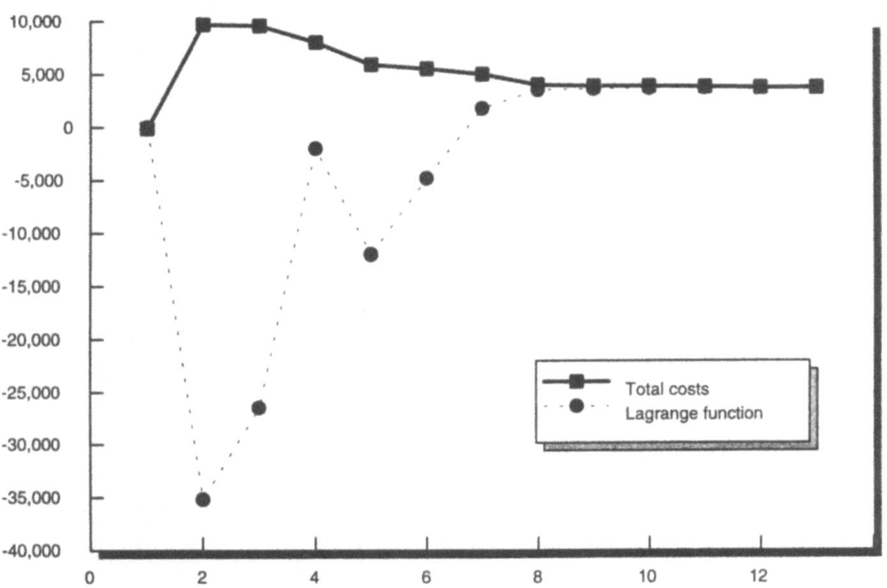

Figure 7.1: Total costs and Lagrange function (10^6 DM/yr) (scenario 1)

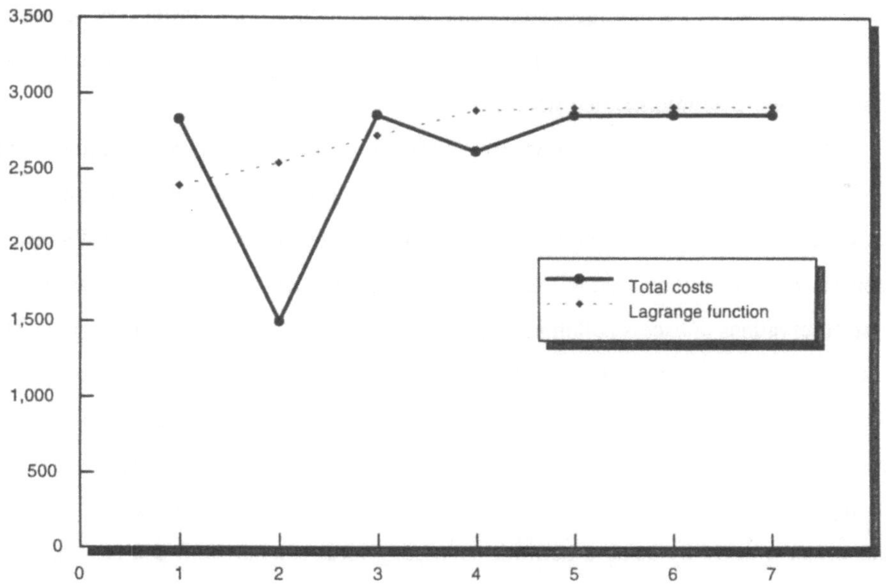

Figure 7.2: Total costs and Lagrange function (10^6 DM/yr) (scenario 2)

Belgium and the Netherlands which have large impacts on the deposition at the receptors located in the Netherlands. Figure 7.2 clearly shows the impact of starting from an initial deposition charge of 100,000 DM/kton NH_3 deposited at each receptor. In this case less iterations would be necessary to approach the cost-minimum solution.

The simulation illustrates that a system of adaptive deposition charges can be formulated that converges to the cost-minimum solution even if the environmental agency has imperfect knowledge on the costs. If some information on possible costs is available to the agency prior to setting the initial charge, the adaptation can proceed faster and lead to less significant fluctuations in costs and emission charge levels than if the agency starts with initial deposition charges equal to zero.

7.7 Conclusion

The aim of this paper was to extend the pricing approach of environmental pollution originally proposed by Baumol and Oates to the case of many receptors when the environmental authority's information on pollution control costs and transfer coefficients is uncertain. Two models to analyze implications of emission and ambient charges policies have been considered: deterministic and stochastic.

For both cases we have shown that, contrary to established opinion, equilibrium vectors of ambient and emission charges that are both cost-effective and environmentally effective can be found by using Walrasian type tâtonnement processes. Just like the Walrasian auctioneer, who uses information on excess demand at given prices, the environmental agency can use observations of excess pollution (concentration) levels to adjust ambient charge levels in a stepwise way, without information on pollution control cost. In the absence of real market signals the artificial market system relies on the ability of a central agency to monitor excess pollution at receptor points, produce "market" signals (charges) and verify emission levels reported by sources. Such a system could find the equilibrium (optimal) set of price signals rather quickly by using a network connecting local computers of sources with the central computer of the agency. The central computer stores the information on the transfer matrix and their probability distributions with a Monte Carlo simulation model enabling to simulate random values of these coefficients. In addition to that the central computer stores actual concentrations and ambient standards. The local computers store the information on corresponding cost functions. By using procedure (7.6) for the deterministic pollution model, or procedure (7.21)–(7.23) for the stochastic model, the agency is able to find the appropriate individual emission charges.

The procedure consists of a learning stage and an implementation stage. The first stage ends with specifying for each emission source the charges and registering of corresponding reported levels of emissions. At the second stage (implementation) the agency verifies actual levels of emissions and monitors depositions at receptors. At this stage the agency has the right to penalize discrepancies between expected and actual emission levels and adjust ambient and emission charges correspondingly. The agency may repeat the adjustment procedure, consisting of stages 1 and 2, each time when actual emissions are expected to change essentially due to technological shifts, aging of control equipment, and so forth. The central computer plays a role as an "artificial world" mixing up emissions from different sources with corresponding random transfer coefficients and providing the agency only information on random excess concentrations at each receptor.

Our conclusion is that it makes no real difference for the feasibility of emission charges whether environmental targets are formulated in terms of emission goals or as a set of ambient concentration standards, even if the central agency has no information on control costs of sources and random transfer coefficients. Under both types of environmental policy objectives an iterative procedure of fixing emission charges can be applied as an instrument that is both cost- and environmentally effective.

Appendix

Asymptotic properties of the adjustment mechanisms

The purpose of this appendix is to provide a formal proof that the adjustment mechanism (7.6) generates the sequence of ambient charges $\lambda^k = \left(\lambda_1^k, \ldots, \lambda_m^k\right)$ $k = 0,1,\ldots$ convergent to a vector λ^* such that derived emission charges $u_i^* \sum_{j=1}^{m} \lambda_j^* h_{ij}$ are cost effective. In other words the decentralized emissions $x_i^*, i = 1,\ldots,n$ minimizing total individual cost $f_i(x_i) + u_i^* x_i$, $x_i \geq 0$ create a solution of the initial problem (7.1)-(7.3). It is important to stress that the convergence of the sequence $\{\lambda^k\}$, $k = 0,1,\ldots$ takes place practically without any specific assumptions on function $f_i(x_i)$. The cost efficiency of the resulting charges follows from the convexity assumptions on costs functions (see corollary below).

Consider the problem (7.1)-(7.3) and sequence $\{\lambda^k\}$, $k = 0,1,\ldots$ defined by eqs. (7.4)-(7.6). Consider the following functions

$$\gamma(\lambda) = \min_{x \geq 0}\left[\sum_{i=1}^{n} f_i(x_i) + \sum_{j=1}^{m} \lambda_j (\sum_{i=1}^{m} (x_i h_{ij} - q_j)\right]$$

The minimization of $\gamma(\lambda)$, $\lambda \geq 0$ is the dual minimax problem to the original pollution control problem (7.1)-(7.3). Under usual assumptions on the convexity of the functions $f_i(x)$ the minimax problem (7.6) is equivalent to the original problem (7.1)-(7.3).

Consider Lagrange function

$$L(x,\lambda) = \sum_{i=1}^{n} f_i(x_i) + \sum_{j=1}^{m} \lambda_j (\sum_{i=1}^{n} x_i h_{ij} - q_j)$$

If $x(\lambda)$ is a solution of the inner subproblem in the definition of the $\gamma(x)$ then also $\gamma(\lambda) = L(x(\lambda),\lambda)$.

Assume that $|\gamma(\lambda)| \neq \infty$ for any $\lambda \geq 0$. Then for any $\lambda \geq 0$ and $\mu \geq 0$

$$\gamma(\mu) - \mu(\lambda) = L(x(\mu),\mu) - L(x(\lambda),\lambda) \leq L(x(\lambda),\mu) - L(x(\lambda),\lambda) = (\Gamma(x(\lambda)),\mu - \lambda), \tag{7.24}$$

for all $\mu \geq 0, \lambda \geq 0$, where $(a,b) = \sum_{j=1}^{m} a_j b_j$ denotes the scalar product of vectors $a = (a_1,\ldots,a_m)$, $b = (b_1,\ldots,b_m)$.

The inequality (7.24) plays essential role in the proof of the convergence. It shows that $\Gamma(x(\lambda))$ is a so-called subgradient of the convex function $\gamma(.)$ at point λ (Clark, 1983) for necessary definition).

Proof. It follows that

$$\|\lambda^* - \lambda^{k+1}\|^2 \leq \|\lambda^* - \lambda^k - \rho_k \Gamma(x^k)\|^2 \leq \|\lambda^* - \lambda^k\|^2 - 2\rho_k(\Gamma(x^k),\lambda^* - \lambda^k) + \rho_k^2 \|\Gamma(x^k)\|^2, \tag{7.25}$$

where $\|\cdot\|$ denotes the Euclidean norm. By eq.(7.13):

$$\gamma(\lambda^*) - \gamma(\lambda^k) \geq (\Gamma(x^k),\lambda^* - \lambda^k) \tag{7.26}$$

and hence

$$(\Gamma(x^k),\lambda^* - \lambda^k) \geq 0,$$

Therefore from the inequality (7.25) we have

$$\|\lambda^* - \lambda^{k+1}\|^2 \leq \|\lambda^* - \lambda^k\|^2 + \rho_k^2 \|\Gamma(x^k)\|^2.$$

From the problem formulation follows that $x_i(\lambda) \leq x_i(0)$ for each i=1,...n and $\lambda \geq 0$. Therefore, the assumption $\|\Gamma(x^k)\| < const$ of the theorem is quite natural. Then for some constant C

$$\|\lambda^* - \lambda^{k+1}\|^2 \leq \|\lambda^* - \lambda^k\|^2 + \rho_k^2 C.$$

Define $\delta_k = \|\lambda^* - \lambda^k\|^2 + C\sum_{s=k}^{\infty} \rho_s^2$.From the above inequality follows that $\delta_{k+1} \leq \delta_k$ and hence the sequence $\{\delta_k\}$ converges. Since $\sum_{k=0}^{\infty} \rho_k^2 < \infty$ the sequence $\{\|\lambda^* - \lambda^k\|^2\}$ is also convergent. From inequality (7.25) and taking into account that $\|\Gamma(x^k)\| < C$ (for some constant C) we have

$$0 \le \|\lambda^* - \lambda^0\|^2 - 2\sum_{k=0}^{\infty} \rho_k \left(\Gamma(x^k), \lambda^* - \lambda^k\right) + C\sum_{k=0}^{\infty} \rho_k^2 .$$

Hence $\sum_{k=0}^{\infty} \rho_k \left(\Gamma(x^k), \lambda^* - \lambda^k\right) < \infty$. Since $\sum_{k=0}^{\infty} \rho_k = \infty$, there exists a sequence $\{\lambda^{k_l}\}$ such that $\left(\Gamma(x^{k_l}), \lambda^* - \lambda^{k_l}\right) \to 0$, $l \to \infty$. Therefore $\gamma(\lambda^{k_l}) \to \gamma(\lambda^*)$, $l \to \infty$.

From this fact and the convergence of the sequence $\left\{\|\lambda^* - \lambda^k\|^2\right\}$ follows the desired result: $\lambda^k \to \lambda^*$, $k \to \infty$.

Corollary. Suppose f_i are strictly convex functions and x_i^k is the solution of the subproblem (7.5) for $\lambda_j = \lambda_j^k$, $j = 1,\dots,m$. Let us show that $x^k \to x^*$, $k \to \infty$

From the definition of $\gamma(\lambda)$ we have $L(x^k, \lambda^k) = \gamma(\lambda^k)$. From the convergence $\lambda^k \to \lambda^*$, $k \to \infty$ and the inequality (7.26) follows that $\gamma(\lambda^k) \to \gamma(\lambda^*)$. Therefore from the duality theory follows that

$$L(x^k, \lambda^k) = \gamma(\lambda^k) \to \gamma(\lambda^*) = L(x^*, \lambda^*).$$

Suppose that x^k does not converge to x^*, that is to say there exists a sequence $x^{k_l} \to \bar{x} \ne x^*$. Then

$$L(x^{k_l}, \lambda^{k_l}) = arg\,\min_{x \ge 0} L(x, \lambda^{k_l}) \to L(\bar{x}, \lambda^*)$$

Since $x^* = arg\,\min_{x \ge 0} L(x, \lambda^*)$ is uniquely defined, $\bar{x} = x^*$. The contradiction completes the proof.

Stochastic Adjustment Procedure

Let us outline a proof of the convergence for the procedure proposed in section 7.5. Although it is described as deterministic adjustment mechanism the essential feature is the use of stochastic estimates (7.18), (7.19). The rigorous analysis requires the stochastic techniques. Estimates of type (7.18), (7.19) allow the regulatory agency to adjust the charges by using only values of excess pollution concentrations (similar to the procedure section 7.3).

In a rather general form the minimisation of the total cost (7.15) subject to the probabilistic constraints (7.16) can be written as the minimisation of a function $g^0(x,y)$ subject to constraints $g^j(x,y) \le 0$, $j=1,\dots,m$, $x \in X$, $y \in Y$. Assume that we are in the situation when this problem is equivalent to the search of a saddle point of the Lagrange function

$$L(x,y,\lambda) = g^0(x,y) + \sum_{j=1}^{m} \lambda_j g^j(x,y),\ x \in X,\ y \in Y,\ \lambda \ge 0.$$

The most restrictive requirement is that $g^0(x,y)$, $g^j(x,y)$ must be convex in (x,y) functions. But it is satisfied in model (7.15), (7.16).

The dual problem then can be written as the maximisation of the function

$$\gamma(\lambda) = \min_x \max_y L(x,y,\lambda) = \min_x L(x, y(x,\lambda), \lambda)$$

Denote $L(x,y(x,\lambda),\lambda) = \varphi(x,\lambda)$ If $(x^*,\lambda^*)$is a saddle point of $\varphi(x,\lambda)$, $x \in X$, $\lambda \ge 0$ then $(x^*, y(x^*,\lambda^*))$ is a solution of the original problem. The procedure (7.21)-(7.23) can be interpreted as the search of a saddle point (x^*,λ^*).

Assume $\varphi_x(x,\lambda)$is the gradient of $\varphi(x,\lambda)$ with respect to x. It exists if $g^0(x,\cdot)$is a strictly convex function for each $x \in X$ what is not restrictive from practical point of view. Then

$$\varphi_x(x,\lambda) = \left[g_x^0(x,y) + \sum_{j=1}^{m} \lambda_j g_x^j(x,y) \right]_{y=y(x,\lambda)}$$

At a current point (x^k, y^k, λ^k), $k = 0,1,\dots$ let $z^k = (z_1^k,\dots,z_n^k)$ be a solution of the problem : minimise

$$(\varphi_x(x^k,\lambda^k), x),\ x \in X.$$

Change now

$$x_i^{k+1} = x_i^k + \gamma_i^k (z_i^k - x_i^k),\ i = 1,\dots,n. \qquad (7.24)$$

A subgradient of function $\varphi(x,\lambda)$ with respect to λ at $(x^k,\lambda^k)$is (see Clarke, 1983).

$$\varphi_\lambda(x^k,\lambda^k) = \left(g^1(x^k,y^k,\lambda^k),\dots,g^m(x^k,y^k,\lambda^k)\right)$$

The procedure (7.21)-(7.23) corresponds to the adjustment (7.24) combined with the adjustment of λ^k in the direction $\varphi_\lambda(x^k,y^k)$ with a step-size multiplier ρ_k. The convergence of such optimisation methods with the use of the statistic estimates of φ_x, φ_λ can be derived from general schemes of stochastic optimisation (see Ermoliev and Wets)). It requires, in particular, that $\gamma^i_k / \rho_k \to 0$, $k \to \infty$, $\sum_k \gamma^i_k = \infty$, $\sum_k \rho_k = \infty$. This conditions are fulfilled when $\rho_k = c/k$, $\gamma^i_k = c_i / k^\varepsilon$, $0 < \varepsilon < 1$ where $c, c_i > 0$ are positive constants.

List of Mathematical Symbols

$x = (x_1, \ldots, x_n)$ - n - dimentional emission vector with components $x_1, \ldots, x_n$ at sources i=1,...n;

$x \in X$ - the vector x belongs to the set X;

η_{ij}, ζ_{ij} - the emission level as a function of random variable v_i, $i = 1, \ldots, n$;

$f_i(x_i)$ - the pollution reduction cost at source i=1,...,n;

$f_i(x_i, v_i)$ - the random pollution reduction cost for emission level x_i and random variable v_i, $i = 1, \ldots, n$;

q_j - the level of ambient standard at receptor $j = 1, \ldots, m$;

λ_j - the shadow price at receptor $j = 1, \ldots, m$;

h_{ij} - deterministic transfer coefficients of emission unit from source i to receptor j;

η_{ij}, ζ_{ij} - random transfer coefficients from i to j;

$\mathrm{Prj}_\Lambda[y]$ - the projection of point y onto the set Λ, or a point in X which minimizes the Euclidean norm (distance) to y;

$\|x\| = \sqrt{x_1^2 + \ldots + x_n^2}$ - the Euclidean norm of the vector x;

$\sum_{i=1}^n x_i = x_1 + \ldots + x_n$;

$(a,b) = \sum_{i=1}^n a_i b_i$ - the scalar product of vectors $a = (a_1, \ldots, a_n)$, $b = (b_1, \ldots, b_n)$;

$\max\{a,b\}$ - is the maximum among numbers a, b, and

$\min\{a,b\}$ - is the minimum among a, b;

$\max_{x \in X} f(x)$ - is the maximum value of $f(x)$ when x varies in a given set X, and

$\arg\max_{x \in X} f(x)$ is a vector, where the maximum value is attained;

if $\min\{a,b\} = a$, then the symbles $a \leq b$ or $b \geq a$ are also used;

$E[\zeta]$ is the mathematical expectation of ζ;

$E[\zeta|A]$ is the conditional mathematical expectation of ζ given events A;

for any two sets A, B $A \supseteq B$ ($A \subseteq B$) means that the set B (A) is included into set A (B);

for any sequence of vectors $a^1, a^2, \ldots, a^k, \ldots$ the limit point is $a = \lim_{k \to \infty} a^k$ or $a^k \to a,\ k \to \infty$.

References

Alcamo, J., R. Shaw and L. Hordijk (Eds.) (1990) *The RAINS model of acidification, science and strategies in Europe.* Kluwer Academic Publishers, Dordrecht/Boston/London.

Baumol, W and W. Oates (1971) The use of standards and prices for protection of the environment. *The Swedish Journal of Economics, 73 (March 1971), pp. 42-54.*

Bohm, P. and C.S. Russel (1985) Comparative analysis of alternative policy instruments. *In: A.V. Kneese and J.L.Sweeney (Eds) Handbook of natural resource and energy economics.* Vol. 1, pp. 395-460. Amsterdam, New York, Oxford: North-Holland.

Clarke, F.H. (1983) *Optimisation and non-smooth analysis.* J.Wiley, New York.

Ellis, J.H. , E.A. McBean and G.J. Fafquhar (1985) Chance-constrained stochastic linesr programming model for acid rain abatement - I. Complete collinearity and noncollinearity. *Atmos. Environ., 19(6), pp. 925-937.*

Ellis J.H., E.A. McBean and G.J. Fafquhar (1986) Chance-constrained stochastic linesr programming model for acid rain abatement - II. Limited collinearity. *Atmos. Environ., 20(3), pp. 501-511.*

Ermoliev, Y. and R. Wets (Eds.) (1988) *Numerical techniques for Stochastic Optimization.* Springer - Verlag, Computational Mathematics.

Ermoliev, Y., G. Klaassen and A. Nentjes (1993) Incomplete Information and the Cost - Efficiency of Ambient Charges. *Forthcoming in Journal of Environmental Economics and Management.*

Fuessle, R.W., E.D. Jr. Brill and J.C. Liebman (1927) Air quality planning: A general chance constrained model. *J.Environ. Engrg., ASCE, 113(1), pp. 106-123.*

Guldmann, J.L. (1986) Interactions between weather stochasticity and the locations of pollution sources and receptors in air quality planning: A chance-constrained approach. *Geographical analysis, 19(3), pp. 198-214.*

Klaassen, G. (1991) *Costs of controlling ammonia emissions in Europe.* Status Report SR-91-03. IIASA, Laxenburg, Austria.

Klein Haneveld, W.K. (1986) *On integrated Chance Constraints. Stochastic Programming, Lecture Notes in Control and Information Sciences* 76. Springer Verlag, New York, pp. 194-209.

Lohani, B.N. and N.C. Thanh (1978) Stochastic programming model for water quality management in a river. *J. Water Pollut. Control Fed., 50(6), pp. 2175-2182.*

Sandnes, H. and H. Styve (1992) *Calculated budgets for airborne acidifying components in Europe, 1985, 1987, 1988, 1989, 1990 and 1991.* EMEP/SC-W report 1/92. Meteorological Synthesizing Center-West, the Norwegian Meteorological Institute, Oslo, Norway.

Tietenberg T.H. (1978) Spatially differentiated air pollutant emission charges: an economic and legal analysis. *Land Economics V54, pp. 265-277.*

VROM (1989) *National Environmental Policy Plan Plus.* Ministry of Housing, Physical Planning and Environment, The Hague, The Netherlands.

8 An Economic Approach to Ozone Abatement in Europe

Inge Mayeres and Stef Proost

Centrum voor Economische Studien

Katholieke Universiteit Leuven

Naamsestraat 69

B-3000 Leuven, Belgium

Abstract

Tropospheric ozone formation is one of the most widespread air pollution problems in Western societies. In the US it has been a matter of concern for 30 years. In Western Europe, public awareness of ozone problems is more recent and has received much less attention than the acidification problem. In this paper we survey the economic aspects of the ozone problem in Europe. Arguably, an economic approach to ozone is still in its infancy in Europe due to the complexity of the ozone formation itself. Ozone formation depends, in a non-linear way, on the presence of two primary pollutants (NO_X and VOC) that are transported over long distances. Therefore, we will begin with a brief introduction to the chemical and physical characteristics of tropospheric ozone before we analyse the basic economics of ozone abatement. For this purpose, we will use an abstract model with growing complexity. The simplest case we will discuss is a model with two primary pollutants without transfrontier transport. It is taken into accoun that No_X, one of the primary pollutants of ozone, also plays a role in the environmental problem of acidification. Therefore the model explicitly considers the side benefits of ozone abatement policies on acidification. The model is then extended by introducing the transfrontier dimension and a member of other features: stochastic pollution, defensive expenditures and the properties of the cost functions. Next, we will survey the European experience in using empirical economic models for the study of ozone. We will examine the estimates of two basic economic components of the ozone problem: damages to

NATO ASI Series, Partnership Sub-Series, 2. Environment – Vol. 14
Economics of Atmospheric Pollution
Edited by Ekko C. van Ierland and Kazimierz Górka

health and agriculture and the emission reduction costs for the two precursors. We will also look at the synthesis of both types of information. Finally, we will conclude that economic modelling efforts are necessary in order to develop a cost-effective and efficient ozone abatement strategy in Europe.

8.1 Ozone Problem: Background

In the stratosphere, ozone protects life on earth against the harmful ultraviolet radiation from the sun. The problem with stratospheric ozone is its depletion by the antropogenic use of CFCs. Unfortunately, ozone is also present in the lower layer of the atmosphere, namely, in the troposphere which stretches from the earth's surface to a height of approximately 10 km.

Ozone is a very strong oxidant and because of this it is harmful for materials, animals, vegetation and people in the troposphere. An increase in long term average ozone concentrations impedes the growth of plants, including trees and agricultural crops. It may also damage some important commercial materials such as auto finishes, rubber and building façades. Exposure to short term elevated concentrations of ozone causes various types of respiratory distress, including increased propensity to asthma attacks, coughing, breathing discomfort and irritation of mucous membranes. Moreover, high short term concentrations also contribute to long-term average concentrations. The WHO has suggested a maximum hourly average ozone concentration of 150 to 200 $\mu g/m^3$. For longer exposure the concentration level is of course lower: 100 to 120 $\mu g/m^3$ for an 8h period. Moreover, a 240 $\mu g/m^3$ maximum hourly concentration should not be exceeded for more than five days a year. According to a European Directive the population should be warned if the hourly average concentration exceeds 180 $\mu g/m^3$.

Ozone is a secondary pollutant. It is formed by complex chemical reactions between volatile organic compounds (VOC)[1], nitrogen oxides (NO_x) and oxygen in the presence of sunlight. Not only the amount of these precursors but also the ratio between them is important. The relationship can be represented graphically by so-called isopleths. These are

---

[1] VOC stands for a group of chemical substances of which the most important are: methane, butune, etheen, isoprene, mehylbenzene, butane.

the loci of combinations of VOC and NO_x concentrations which give rise to the same ozone concentration level. Figure 8.1 presents isopleths which are derived from laboratory experiments. To illustrate the complexity of the relationship between ozone and its precursors we assume that we start from point A on the 240 μg/m^3 isopleth. This corresponds with a NO_x and VOC concentration of approximately 250 μg/m^3. In point B the ozone concentration is the same but the NO_x concentration is much lower, 30 μg/m^3. This means that, at constant VOC concentration, the reduction in the NO_x concentration by 88% has no effect on ozone. Moreover, if NO_x is reduced by, for example, only 50%, the resulting ozone concentration is higher than before (appr. 400 μg/m^3). The isopleths show us that the same ozone concentration can be reached under different scenarios, which will be important if one wants to determine optimal ozone control strategies. Although the isopleths are derived on the basis of laboratory experiments, in real conditions similar isopleths will also be valid. But the actual form of the isopleths can differ from region to region because of geographical or climatological reasons, the composition of the VOC, etc. Thus, the same reduction strategy can have different effects on the ozone levels in different regions. This is illustrated in table 8.1. In this table peak ozone concentrations are given for the different European countries. Most countries have peak ozone concentrations that exceed the current WHO standards. The table gives the percentage reductions in ozone for three emission scenarios: 50% NO_x reduction, a 50% SO_2 reduction and the combination of both abatement scenarios. All abatement scenarios imply a uniform reduction effort in all European countries. The ozone results illustrate the non-linear character of the ozone isopleths and their regional diversity: a substantial decrease in NO_x emissions (and NO_x concentrations) leads to an ozone increase in four countries. Some countries are more sensitive to NO_x reductions while others need VOC reductions to improve their peak ozone concentrations.

The tropospheric ozone problem is a regional problem. It is not a local problem because the pollutants that cause ozone and ozone itself are transported over several hundreds of kilometres and because economies of scale in some of the abatement strategies can only be realised with widespread adoption.

The precursors of ozone are VOC and NO_x. In the European Community the transportation sector is responsible for 31% and 54% of the annual antropogenic emissions of VOC and NO_x respectively. Other sources of VOC emissions in Europe are solvent evaporation (20%), industrial combustion and processes (5%) and other sources (44%). For

NO_x power utilities contribute 24 % and industrial combustion and processes 16% (Eurostat, 1991). For VOC non-antropogenic sources also contribute to emissions, especially in the spring and summer.

8.2 Basic ozone economics

8.2.1 Model without transboundary effects

In the first model things are kept simple. We consider the case of a homogeneous region where the emissions of each pollutant lead to uniform concentrations over the territory, and we assume that the climatic conditions and emissions are also uniform within the period considered. We consider two pollutants: NO_x and VOC. Both contribute to the tropospheric ozone problem. However, NO_x is also one of the sources of the acid rain problem, which will be taken into account in the model as well. $X_n(0)$ and $X_h(0)$ give the total emissions of NO_x and VOC in the reference state. Q^0 is the corresponding level of ozone and for given emissions of SO_2, P^0 gives the level of acidification. Furthermore, we define x_n and x_h as the abatement levels of NO_x and VOC starting at the reference level. $\bar{x}_s$ is the abatement level of SO_2 and is exogenous. The total abatement cost is given by the continuous cost function $C(x_n,x_h)$. The ozone concentration is determined by the continuous ozone formation function $Q(x_n,x_h)$ and depends on the abatement of NO_x and VOC. The continuous function $P(x_n, \bar{x}_s)$ gives the acidification level for a given level of SO_2 abatement.

The control problem then is to choose x_n and x_h such that the sum of the damages due to both ozone and acidification and the abatement costs are minimised. It can be represented as follows:

$$\underset{x_n,\, x_h}{Min} \left[D\left(Q\left(x_n, x_h\right)\right) + P\left(x_n, \bar{x}_s\right) + C\left(x_n, x_h\right)\right] \qquad x_n, x_h \geq 0 \tag{8.1}$$

The solution to this problem depends on the shape of the cost and the damage functions. The shape of the cost function is typically assumed to be convex. For an additive cost function,

Table 8.1: Percentage reductions in calculated maximum ozone when either 50% No_x, 50% VOC, or 50% No_x + VOC emission control is applied using the EMEP model

	Base maximum ozone concentration ($\mu g/m^3$)	Emission scenario		
		50% NO_x	50% VOC	50% NO_x + VOC
Albania	282	17	12	26
Austria	266	8	20	24
Belgium	252	9	15	23
Bulgaria	212	12	26	27
Czechoslovakia	252	11	20	25
Denmark	236	-7	18	10
Finland	234	17	5	23
France	248	12	7	25
G.D.R.	252	11	20	25
F.R.G.	248	-2	14	15
Greece	230	17	3	20
Hungary	266	8	21	24
Ireland	212	-1	20	21
Italy	288	14	13	25
Luxembourg	248	12	27	25
Netherlands	252	9	15	23
Norway	228	16	1	24
Poland	226	-2	17	15
Portugal	186	16	6	19
Romania	186	11	16	22
Spain	328	21	6	24
Sweden	274	11	17	23
Switzerland	222	12	13	24
Turkey	184	12	12	20
U.S.S.R.	238	19	1	27
U.K.	230	0	26	22
Yugoslavia	266	8	18	24

Source: Simpson (1991)

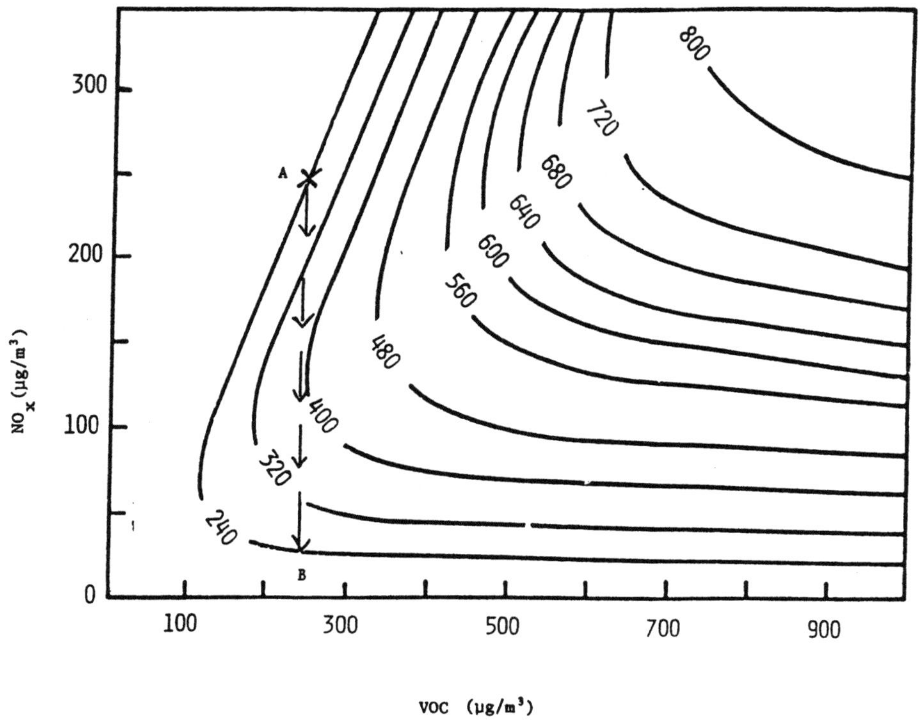

Figure 8.1: Ozone chemistry

Source: Proost, Vinckier, Mayeres, Nemery (1995)

rising marginal abatement costs are the rule, and this is sufficient to have C convex and the isocost curves as quasiconvex.

Less is known about the environmental damage curves. We can assume that the marginal damage of ozone is increasing in the ozone level:

$$\frac{\partial^2 D}{\partial Q^2} > 0 \tag{8.2}$$

and that the marginal acidification damage is decreasing in the NO_x abatement:

$$\frac{\partial^2 P}{\partial x_n^2} < 0 \tag{8.3}$$

If the ozone formation function Q were linear in the abatement of both precursors, the previous assumptions would guarantee that (8.1) is a well-behaved optimisation problem with decreasing marginal benefits for ozone and acidification at increasing marginal costs. This property holds for the optimal control of both precursors. The quasi-convexity of the cost function guarantees the existence of an interior solution.

We know that the ozone formation function Q is non-linear in the abatement of both precursors and probably quasi-convex in the (x_n, x_h) space.

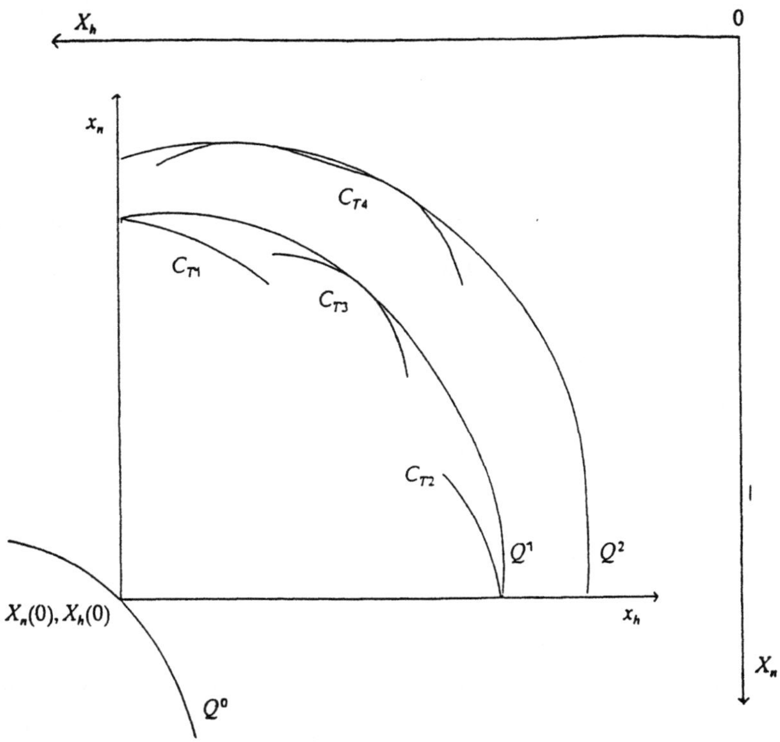

Figure 8.2: Non-convex iso-ozone curves and corner solutions

Source: Mayeres, Miltz, Proost (1992)

We know from optimisation theory that in this case, the minimum problem (8.1) can, but does not have to deliver corner solutions (Luenberger, 1970). This is illustrated in figure 8.2. Starting from the reference situation $X_n(0)$, $X_h(0)$, one defines isocost curves in the (x_n, x_h)-space that can have different forms: C_{T1}, C_{T2}, etc... One can also define iso-damage curves for ozone that correspond to isopleths Q^1, Q^2 etc... The first problem is to reach the best isopleth and this with a given total cost. This may require a corner solution: compare the optimal mix for a cost function C_{T1} and for a cost function C_{T2}. Next, one has to select the isopleth for which the marginal reduction in ozone damage equals the marginal cost of ozone reduction.

The first order necessary conditions for an optimum are given by:

$$\frac{\partial D}{\partial Q} Q_n \geq -(C_n + P_n) \qquad x_n \geq 0 \;\; and \;\; x_n\left(\frac{\partial D}{\partial Q} Q_n + C_n + P_n\right) = 0$$
$$\frac{\partial D}{\partial Q} Q_h \geq -C_h \qquad x_h \geq 0 \;\; and \;\; x_h\left(\frac{\partial D}{\partial Q} Q_h + C_h\right) = 0 \tag{8.4}$$

C_n, C_h, Q_n, Q_h and P_n denote the partial derivatives of C, Q and P. If we have an interior solution, these lead to the familiar finding that VOC and NO_X should be reduced to the point where marginal costs of ozone reduction, through reductions in the precursor, equal the marginal benefits. The marginal costs of NO_X reductions are corrected by the fact that NO_X reductions have a side benefit in terms of their effect on the acidification problem.

Moreover, the first order conditions can be combined to give the following expression:

$$\frac{C_n + P_n}{C_h} = \frac{Q_n}{Q_h} \tag{8.5}$$

The optimal strategy thus requires that the marginal rate of substitution between NO_X and VOC that keeps peak ozone concentrations unchanged equals the marginal rate of substitution between NO_X and VOC that keeps total costs unchanged. The costs of reducing NO_X emissions are corrected by the benefit that NO_X reductions imply in terms of the acid rain problem. If both pollutants are controlled, the reduction in ozone per additional unit of money spent on NO_X and VOC should be the same[2].

---

[2] Repetto (1987) was the first to present the ozone problem in terms of isocost and iso-ozone curves. The side-benefits on acidification were only invoked ex-post in the sense that in the case of indeterminancy in the choice between NO_x and VOC abatement, one can as well give a preference to No_x abatement. Mayeres, Miltz and Proost (1993) have included the side-benefits of NO_x abatement explicitly in the analysis.

8.2.2 Model with transboundary effects

Ozone pollution has a strong transboundary component. In order to analyse the impact of this dimension we use a setting with only 2 countries (denoted by superscripts H (home) and F (foreign) respectively). Three matrices 2x3 transport coefficients are used: t^{HH}, t^{HF} and t^{HR} represent, respectively, the proportion of the emissions of the home country remaining in the country, the proportion transported abroad and finally the remaining portion that is harmless. The sum of the three coefficients equals 1. We have one matrix for VOC (index h), one for NO_x contributing to ozone (index n) and one for NO_x contributing to acidification (index r). The reduction in the concentration of precursors in country H for pollutant n now equals a weighted sum of efforts at home and abroad:

$$t_n^{HH} x_n^H + t_n^{FH} x_n^F$$

The formulation of the ozone control problem depends on the international institutional context. We will briefly discuss three alternative settings: a non-cooperative equilibrium; a fully cooperative solution; and a federal type of setting.

In the *Nash non-cooperative solution* every country fully controls its own emissions and takes the abatement decisions of the other countries as given. Each country minimises total damages and costs:

$$\begin{aligned} \underset{x_n,\, x_h}{Min} \;\; & D^H\left[Q^H\left(t_n^{HH} x_n^H + t_n^{FH} x_n^F,\, t_h^{HH} x_h^H + t_h^{FH} x_h^F\right)\right] \\ & + P^H\left(t_r^{HH} x_n^H + t_r^{FH} x_n^F\right) + C^H\left(x_n, x_h\right) \end{aligned} \tag{8.5}$$

In the case of an interior solution[3], we have the following definition of the reaction functions:

$$C_n^H = t_n^{HH} \frac{\partial D^H}{\partial Q^H} Q_n^H + t_r^{HH} P_n^H \tag{8.6}$$

[3] Note that when there is no interior solution, the reaction functions will be discontinuous and the existence of a Nash-equilibrium is no longer guaranteed.

$$C_h^H = t_h^{HH} \frac{\partial D^H}{\partial Q^H} Q_h^H \tag{8.7}$$

The reaction functions (8.6) and (8.7) are a function of the other countries' emissions via the ozone formation and acidification functions[4].

The reaction functions clearly illustrate that each country takes into account only the marginal damages in its own country and tends to concentrate more on the pollutant that is less easily exported.

In the *fully cooperative solution* the weighted sum of total damages and total costs of all countries is optimised. The welfare weights w^H, w^F will equal 1 if side payments between the countries are possible. When side payments are not possible, the welfare weights have to be chosen such that the participation constraints are verified for all countries[5]. The participation constraints ensure that each country participating in the cooperative solution is better off than in the non-cooperative solution. The optimal abatement levels now correspond to (in the case of an interior solution):

$$w^H C_n^H = w^H t_n^{HH} \frac{\partial D^H}{\partial Q^H} Q_n^H + w^H t_r^{HH} P_n^H + w^F t_n^{HF} \frac{\partial D^F}{\partial Q^H} Q_n^F + w^F t_r^{HF} P_n^F \tag{8.8}$$

$$w^H C_h^H = w^H t_h^{HH} \frac{\partial D^H}{\partial Q^H} Q_h^H + w^F t_h^{HF} \frac{\partial D^F}{\partial Q^H} Q_h^F \tag{8.9}$$

In the right-hand sides of (8.8) and (8.9), we see that damages abroad are fully taken into account now. The magnitude of the welfare gain of cooperation with side payments can be approached intuitively by taking a very simplified case with only 1 pollutant, identical countries, constant marginal benefits b, and a marginal cost function equal to cx^H. In that case, the welfare gain equals:

---

[4] When there are economies of scale in abatement decisions (e.g., European-wide adoption of a catalytic converter for cars), there are also interdependencies via the cost function.

[5] See Eyckmans, Proost, Schokkaert (1994) for an application of this technique to the climate change problem.

$$0.5\frac{\left(bt^{HF}\right)^2}{c}$$

This is the familiar welfare triangle: the higher the exported share of pollution (the smaller the geographic size of the country) and the higher the marginal benefit, the higher is the gain of cooperation. The gain of cooperation will be much smaller when the marginal abatement cost curves have a high slope (c). The intuition is that the extra abatement effort warranted by the exported pollution damage, comes at a high cost such that the net welfare gain of expanding abatement is small.

When side-payments are not available, voluntary collaboration can be achieved only at an efficiency cost. In order to fulfil the participation constraint in (8.8) and in (8.9), it is necessary to have welfare weights w which are higher than those for the countries that are the net exporters of pollution, or low cost abaters, and this causes efficiency losses[6].

Achieving the cooperative solution through voluntary cooperation raises not only the problem of participation but also of preference revelation. Countries with low abatement costs and high damages have an incentive to misrepresent their preferences so as to minimise their contribution in the total efforts. There exist incentive mechanisms that guarantee participation and truthful revelation of preferences. Due to their complexity though, they have largely not been applied as yet[7].

In a *federal solution*, a central government can impose, to some extent, abatement efforts on its member states (e.g., the European Union). The federal government is not bound by the participation constraints of its member states as it has explicitly received the power to enforce solutions upon its member states. Furthermore, it can compensate disequilibria in benefits and costs through other instruments. It faces, however some "uniformity" constraints[8] that limit the flexibility in the federal regulation. With the help of a simple model, Braden & Proost (1995) analyse the potential use, by the federal government, of uniform and maximum regulations on emissions of precursors and on ambient concentration of ozone. They use a Stackelberg setting in which the federal government looks for the best regulation of state

---

[6] The magnitude of the efficiency losses caused by the participation constraint was found to be rather small in the case of global warning (see Eyckmans, Proost, Schokkaert, 1994).

[7] Kaitala, Mäler, Tulkens (1995), applied this technique to the acidification problem.

[8] These "uniformity" constraints stem from the constitution. This is a "cooperation" constraint that limits the power of the federal government to "exploit" a minority of states.

instruments, taking into account the fact that the state governments pursue their own regional objective. It is assumed that the federal government knows the preferences of the states[9]. Their results depend strongly on the asymmetry present in the damages, costs and transport coefficients of the member states. The simple models are useful to understand the inefficiency of the maximum ozone ambient regulation used in the US in the past, and the maximum emission regulations that are more familiar in the EU.

8.2.3 Introducing other Problem Dimensions

A proper model for the study of ozone economics should take into account three more features: stochastic pollution, defensive expenditures and the properties of the cost function. The peak ozone levels are, certainly in Europe, *a stochastic problem.* Ozone peaks require a combination of sun and wind bringing the "bad" concentration of the different pollutants. A stochastic pollution problem requires a good knowledge of the slope of the marginal damage function: how much more damaging is a peak ozone problem compared with an average ozone problem? The fact that ozone pollution is limited to a few days (weeks) a year also opens up new abatement possibilities that are only effective during these days. One of these possibilities is the restriction, rather than the "cleaning", of certain economic activities: for example, reducing car traffic during the sunny periods rather than imposing a low emission standard on all cars. Another possibility is a better organisation of *defensive measures.* The health damage caused by ozone can be strongly limited by restricting the outdoor activities of some population groups. Investment in a good warning system can also be a cost-effective public investment.

Finally a good ozone economics model should incorporate two improvements in the representation of the *cost function.* The first is to allow for cost functions that are non-separable in the different pollutants. Installing pollution abatement equipment carries a high fixed cost and can be useful at a small extra cost for several pollutants. The three-way catalytic converter in cars reduces CO, NO_X and VOC at the same time. The second

[9] When the federal government does not know the state preferences, the federal government has to propose regulation schemes that induce a thruthfull revelation of state preferences. A. Ulph (1995) is one of the first to use this more complex setting for federal environmental problems.

improvement is the consideration of economics of scale in the production of abatement equipment. Taking again the example of cars, Harrington, McConnell & Walls (1995) stated that the average cost of abatement equipment for cars would be only 50 to 90% of Californian costs if the whole US, rather than California alone, enforced the installation of this abatement equipment.

8.2.4 Using the economic framework

In order to carry out the analysis described in the previous section, one needs information about the cost functions for reducing NO_x and VOC emissions and about the benefits of ozone reduction. This information has to be gathered for each European country. In addition, one needs an atmospheric model which relates VOC and NO_x emissions in each country to the levels of ozone concentration in all countries. Unfortunately, not all of this information is available at the moment. In Europe, most of the research concerns the construction of cost curves for the reduction of pollutants. One of the most successful undertakings has been the RAINS-model[10], which contains cost functions for the reduction of SO_2 in all European countries, together with an atmospheric transport model and an index of acidification damage in each European country. As will be indicated later, only part of the information needed on the benefits of ozone reduction and the transformation and transportation of ozone and its precursors is currently available in Europe.

In this section we present an illustration from Belgium[11]. We discuss the best international negotiation strategy for ozone reduction in Belgium given that the emission reduction strategies are limited to uniform emission reductions in all European countries. This kind of strategy is probably not optimal, but it allows us to present the type of information that needs to be gathered in order to analyse the economics of the ozone problem in Europe.

[10] RAINS stands for Regional Acidification Information and Simulation model. For a description see Alcamo e.a. (1991).

[11] Van Ierland (1995) treats ozone reduction and reduced acidification as a side-benefit of global warming policy and considers the case of two countries: West-Germany and The Netherlands.

A. *The effects of VOC and NO_x reduction on ozone levels*

In order to know the benefits of VOC and NO_x reduction in terms of their effect on ozone, one needs to go through three steps, each of which will be discussed briefly. First, the relationship between ozone concentrations in Belgium and Belgian and foreign VOC and NO_x emissions needs to be determined. This can be done with the help of atmospheric models. At the moment, only limited information exists on this point. However, studies of EMEP (Simpson (1991)) and De Leeuw & Van Rheineck Leyssius (1991) already give an idea of the relationship between European VOC and NO_x reductions and the ozone concentration in Belgium. On the basis of this information, we can construct isoreduction lines. These are the loci of combinations of European VOC and NO_x reduction scenarios which give rise to the same reduction in ozone concentration. The isoreduction lines are presented in figure 8.3. These curves are labelled -2.9, -6.7, -14, -15, etc.. The labels present the percentage reduction of ozone levels in Belgium that can be achieved for different combinations of uniform European VOC and NO_x-reductions. They are the mirror image of the isoreduction lines of Figure 8.1 and are constructed on the basis of the findings of De Leeuw & Van Rheineck Leyssius (1991). It should be noted that while there is an uncertainty concerning the exact level of the ozone reduction (different levels are found in, e.g., Simpson (1991)), there is more certainty about the form of the isoreduction lines. The isoreduction lines will not have the same form for all European countries. Those presented here are typical for the situation in countries in North-western Europe (inter alia Belgium, Luxemburg, and the Netherlands) where it is possible that a reduction in NO_x actually increases ozone concentrations, an effect which is less likely to occur in the Southern European countries. The full European model that is needed would contain ozone reduction lines for each country as a function of the abatement efforts of NO_x and VOC for all other countries. The EMEP-model (Simpson (1991)) can generate this information.

B. *The benefits of an ozone reduction*

The second step consists in assigning a monetary value to changes in the ozone concentration. Different techniques are available to do this. They include, inter alia, the revealed preference methods and stated preference method. We will primarily use a production function approach that computes the cost savings in the production of health and agricultural products that are generated by lower ozone levels. Ozone has an effect on health, agriculture, vegetation,

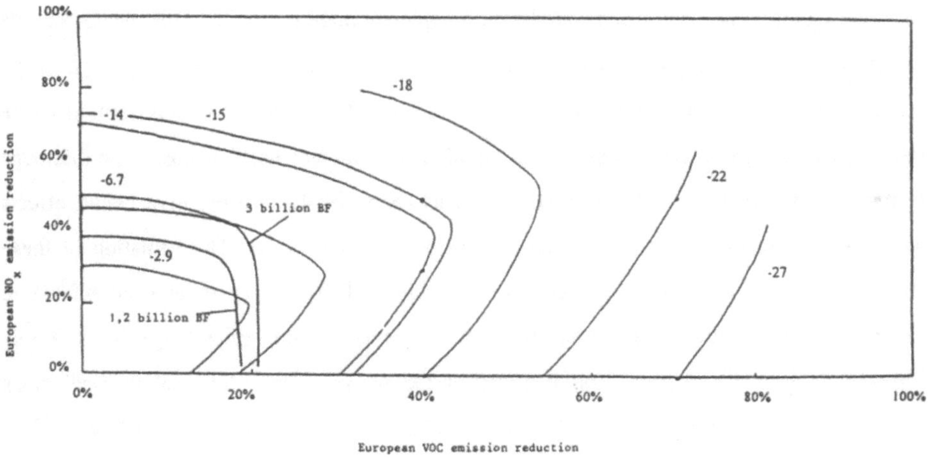

Figure 8.3: Selection of the optimal Belgian negotiation strategy

Source: Proost, Vinckier, Mayeres, Nemery (1995)

materials and visibility. Most work has been done on finding dose-response relationships for the first three elements, to which we will also limit our discussion. The number of European studies on this topic is very limited[12]. Therefore, one has to base the calculations on the findings of American studies. However, it is uncertain to what extent these findings can be transferred to the European situation. Thus, there is a great need for European studies in this field.

One of the most extensive surveys of the valuation of air pollution can be found in the ExternE project of the European Commission (Metroeconomica (1994)). The aim of this research project was to estimate the external costs of power plants. The application in Maes, Proost & Schokkaert (1994) serves however our purpose better. On the basis of American studies the authors calculate the benefits for Belgium of an ozone reduction by 10% (excess ozone and average ozone levels). Since no dose-response relationships are available for Belgium, the relationships used were those estimated for the US. The dose-response relationships relate the number of days of respiratory problems to excess ozone, accounting

[12] There are a few interesting studies on the health impacts of ozone (Hoek e.a. (1994)) but not on the economic valuation of these effects.

for other factors. The relationships are based on epidemiological studies. This means that the health problems of a representative sample of the population are related to observed ozone concentrations and other potentially explanatory factors. These kinds of studies implicitly take into account the defensive actions of the population to protect itself against ozone damage (e.g. by staying inside). In Maes, Proost & Schokkaert (1994), two negative health effects were considered: days with rest in bed and restricted activity days. The valuation of these effects consists of three parts: production loss, additional medical costs, and the subjective valuation of the distress. The production loss is valued at the gross wage because this corresponds with the marginal productivity of the worker. The additional medical costs consist of the transport costs to the hospital or doctor and the total costs of medical treatment. Finally, the subjective valuation of being ill is taken into account. This is determined by asking a representative sample of people how much they are willing to pay to avoid a day with asthma, headaches etc. The production loss and the additional medical costs account for 70% of the total costs. The findings for the US were extrapolated to Belgium taking into account the difference in the extent of the ozone problem and the gross wages. It is found that a reduction of excess ozone by 10% would have a benefit of approximately 3.67 billion BF[13].

In order to estimate the damage to agriculture, the same two steps have to be undertaken. Maes, Proost & Schokkaert (1994) based their studies on American dose-response relationships. The extra yield due to the reduction in ozone is valued in terms of the additional gross profit (change in sales income minus costs of extra inputs) in which it results. The gross profit is calculated on the basis of world prices, i.e., prices without subsidies. This is an important aspect. While the income for the agricultural sector increases at subsidised prices, the benefit for society as a whole is lower, namely, the increase in income measured in world prices. Due to data problems the two steps could not be carried out for Belgium. Therefore, the American benefits for agriculture were extrapolated to Belgium. In this process, account was taken of the lesser relative importance of agriculture and the lower ozone concentration in Belgium during the growing season. For Belgium the benefits of the ozone reduction would amount to 470 million BF. The damage to non-agricultural vegetation would amount to 250 million BF.

[13] 1 BF is approximatively 0.025 ECU.

Summing up the health, agricultural and vegetation effects of the 10% ozone reduction, we arrive at a value of 4. 4 billion BF. The value of a 7% ozone reduction would be somewhat higher than 3 billion BF. This information is not yet sufficient for the exercise we want to carry out. We require marginal damage curves and not only the total damage at a given level of ozone reduction.

C. The cost of NO_x and VOC abatement

The third type of information which is needed for the analysis is cost information. Less problems exist in this field. In the application for Belgium the abatement cost data are based on estimates found in the literature (see Mayeres, Miltz & Proost, 1993). This allows us to construct marginal cost curves. The isocost curves are derived on the basis of NO_x and VOC abatement cost functions. Account is taken of a credit for NO_x abatement as it also contributes to a lower acidification level. The credit is computed by taking, for each unit of NO_x reduction, the avoided costs of the equivalent SO_2 reduction. Introducing the NO_x credit does not change the general shape of the isocost curve but makes it less quasiconvex. The resulting isocost curves are presented in figure 8.3 and are labelled in billion BF.

D. Selection of an abatement strategy

The confrontation of the isocost curves and the isoreduction lines indicates what is the most cost effective strategy to reduce ozone concentration. For Belgium what is the optimal mix of European NO_x and VOC reductions given that one wants to reduce ozone by a certain percentage? It is clear from figure 8.3, that the best strategy consists of reducing the VOC emissions because it is on the x-axis that each isoreduction line touches the lowest isocost line.

It is more difficult to find the optimal level of ozone reduction for Belgium. From figure 3, we see that the lowest cost for Belgium for realising a 7% ozone reduction, through a uniform European abatement effort, would be approximately 3 billion BF per year, while the benefits were computed to be somewhat higher than 3 billion BF per year. However, from this one cannot conclude that the minimum level of ozone abatement should be 7% as the difference between total benefit and total costs could be higher for lower ozone reduction levels. If on the contrary, the benefits of an additional reduction are higher than the associated costs, it would be optimal to reduce ozone by more than 7%.

This empirical illustration has been limited to the selection of an optimal negotiation position for Belgium wherein only uniform European strategies were accepted. The real problem is much more complex: to select, simultaneously for all countries, abatement levels for NO_X and VOC (and implicitly ozone levels) such that the difference between total benefits and total costs is maximised, while maintaining the participation of all countries in the agreement.

8.3 Conclusions

In this survey we have identified research priorities for an economic approach to ozone abatement in Europe. Of the three necessary elements, ozone damage estimates, ozone formation and transport functions, and estimates of abatement costs, the latter two are more or less available. Several atmospheric ozone models exist (on a European scale) that could be used to generate isoreduction lines for the different European countries. Estimates of abatement costs for NO_X and VOC exist in most countries as this information is necessary for a cost-effective reduction strategy for both precursors. The most important gaps exist in the estimation of the marginal benefits of ozone reduction and in the assembly of the three necessary elements. These gaps can be filled by applying available methodologies. For too long, European ozone policies have been based on intuitively appealing, but counterproductive and costly, abatement strategies that impose a uniform reduction of all primary pollutants in all countries.

References

Alcamo, J., R. Shaw and L. Hordijk (Eds.) (1990) *The RAINS model of Acidification, Science and Strategies in Europe.* Kluwer Academic Publishers, Dordrecht.

Braden, J. and S. Proost (1995) Economic Assessment of Policies for Combating Tropospheric Ozone in Europe and the US. In: J. Braden, e.a., *Environmental Policy with Economic and Political Integration: the European Union and the United States*, Edward Elgar Publishers.

Eurostat (1991) *Environmental Statistics.* European Communities, Luxembourg.

Eyckmans J., S. Proost and E. Schokkaert (1994) A Comparison of Three International Agreements on Carbon Emission Abatement. In: E. van Ierland (Ed.), *International Environmental Economics*, Elsevier, Amsterdam.

Harrington, W., V. McConnell and M. Walls (1995) "Who is in the Driver's Seat? Mobile Source Policy in the U.S. Federal system" Paper presented at the symposium on "Economic Aspects of Environmental Policy Making in a Federation", Leuven, June 1995.

Hoek, G., P. Fischer, B. Brunekreef, E. Lebret, P. Hofschreuder and M.G. Mennen (1993) Acute Effects of Ambient Ozone on Pulmonary Function in the Netherlands. *American Review of Respiratory Diseases, 147, pp. 111-117.*

Ierland van E., (1995) "National and International Policies for Global Warming: Side Benefits for Acidification and Tropospheric Ozone". Paper presented at the symposium on "Economic Aspects of Environmental Policy Making in a Federation", Leuven, June 1995.

Kaitala, V., K.-G. Maler and H. Tulkens (1993) The Acid Rain Game as a Resource Allocation Process with an Application to the International Cooperation among Finland, Russia and Estonia. *Scandinavian Journal of Economics, 97(2), pp. 325-343.*

Leeuw de, F.A. and H.J. Van Rheineck Leyssius (1991) Calculation of Long Term Averaged and Episodic Concentrations for the Netherlands. *Atmospheric Environment 25 A, pp. 1809-1818.*

Luenberger, D.G., (1972), *Introduction to Linear and non-linear programming.* Addison-Wesley, New York.

Maes, J., S. Proost and E. Schokkaert (1994) Economische waardering van milieuschade. In: A. Verbruggen (Red.), Hoofdstuk V:2C uit: *Milieu- en natuurrapport Vlaanderen*, Garant (in Dutch).

Mayeres, I., D. Miltz and S. Proost (1993) The Geneva Hydrocarbon Protocol: Economic Insights from a Belgian Perspective. *Environmental and Resource Economics, 3, pp.107-127.*

Metroeconomica (1994) *Economic Valuation - Externalities of Fuel Cycles - ExternE project.* Commission of the European Communities, DG XII.

Proost, S., C. Vincier, I. Mayeres and B. Nemery (1995) Ozon - eerst denken dan doen. *Leuvense Economische Standpunten* 80, C.E.S.-K.U.Leuven (in Dutch).

Repetto, R., (1987) The Policy implications of Non-Convex Environmental Damages: A Smog Control Case Study. *Journal of Environmental Economics and Management 14, pp.13-29.*

Simpson, D., (1991) The EMEP MSC-W Photo-oxidant-Model. In: T. Iversen (Ed.), *Comparison of Three Models for Long Term Photochemical Oxidants in Europe*, The Norwegian Meteorological Institute, Norway.

Simpson, D., (1992) *Long Period Modelling of Phototechnical Oxidants in Europe, A) Hydrocarbon reactivity and Ozone formation in Europe, B) On the linearity of Country Ozone Calculations in Europe* (EMEP/MSC-W, Oslo), EMEP/MSC-W Report 1/92.

Ulph, A., (1995) "International Environmental Regulation when National Governments Act Strategically", paper presented at the symposium on "Economic Aspects of Environmental Policy Making in a Federation", Leuven, June 1995.

INDEX

F

G

P

Q

R

The Partnership Sub-Series incorporates activities undertaken in collaboration with NATO's Cooperation Partners, the countries of the CIS and Central and Eastern Europe, in Priority Areas of concern to those countries.

The volumes published as a result of these activities are:

Vol. 1: **Clean-up of Former Soviet Military Installations.** Edited by R. C. Herndon, P. I. Richter, J. E. Moerlins, J. M. Kuperberg, and I. L. Biczó. 1995

Vol. 2: **Cleaner Technologies and Cleaner Products for Sustainable Development.** Edited by H. M. Freeman, Z. Puskas, and R. Olbina. 1995

Vol. 3: **Remediation and Management of Degraded River Basins.** Edited by V. Novotny and L. Somlyódy. 1995

Vol. 4: **Earthquakes Induced by Underground Nuclear Explosions.** Edited by R. Console and A. Nikolaev. 1995

Vol. 5: **Transportation Infrastructure.** Edited by R. M. Gutkowski and J. Kmita. 1996

Vol. 6: **Sustainable Development of the Lake Baikal Region.** Edited by V. A. Koptyug and M. Uppenbrink. 1996

Vol. 7: **Transboundary Water Resources Management.** Edited by J. Ganoulis, L. Duckstein, P. Literathy, and I. Bogardi. 1996

Vol. 8: **Urban Air Pollution.** Edited by I. Allegrini and F. De Santis. 1996

Vol. 9: **Advances in Groundwater Pollution Control and Remediation.** Edited by M. M. Aral (Kluwer Academic Publishers). 1996

Vol. 10: **Assessing the Risks of Nuclear and Chemical Contamination in the Former Soviet Union.** Edited by E. S. Kirk (Kluwer Academic Publishers). 1996

Vol. 11: **Ventilation and Indoor Air Quality in Hospitals.** Edited by M. Maroni (Kluwer Academic Publishers). 1996

Vol. 12: **The Aral Sea Basin.** Edited by P. P. Micklin and W. D. Williams. 1996

Vol. 13: **Radioecology and the Restoration of Radioactive-Contaminated Sites.** Edited by F. F. Luykx and M. J. Frissel (Kluwer Academic Publishers). 1996

Vol. 14: **Economics of Atmospheric Pollution.** Edited by E. C. van Ierland and K. Górka. 1996

Vol. 15: **Water Supply Systems.** Edited by C. Maksimović, F. Calomino, and J. Snoxell. 1996

Vol. 16: **Bioindicator Systems for Soil Pollution.** Edited by N. M. van Straalen and D. A. Krivolutsky (Kluwer Academic Publishers). 1996

Vol. 17: **Remediation of Soil and Groundwater – Opportunities in Eastern Europe.** Edited by E. A. McBean, J. Balek, and B. Clegg (Kluwer Academic Publishers). 1996

Vol. 18: **Environmental Engineering and Pollution Prevention – European Network on Excellence and Partnership.** Edited by J. Wotte, W. A. Halang, and B. J. Kraemer (Kluwer Academic Publishers). 1996

Vol. 19: **East-West Life Expectancy Gap in Europe – Environmental and Non-Environmental Determinants.** Edited by C. Hertzman, S. Kelly, and M. Bobak (Kluwer Academic Publishers). 1996

Springer-Verlag and the Environment

We at Springer-Verlag firmly believe that an international science publisher has a special obligation to the environment, and our corporate policies consistently reflect this conviction.

We also expect our business partners – paper mills, printers, packaging manufacturers, etc. – to commit themselves to using environmentally friendly materials and production processes.

The paper in this book is made from low- or no-chlorine pulp and is acid free, in conformance with international standards for paper permanency.

MIX
Papier aus verantwortungsvollen Quellen
Paper from responsible sources
FSC® C105338

If you have any concerns about our products,
you can contact us on
ProductSafety@springernature.com

In case Publisher is established outside the EU,
the EU authorized representative is:
Springer Nature Customer Service Center GmbH
Europaplatz 3, 69115 Heidelberg, Germany

Printed by Libri Plureos GmbH
in Hamburg, Germany